Leckie × Leckie
Scotland's leading educational publishers

National 5
LIFeSKILLS MATHS
STUDENT BOOK

N5 LIFeSKILLS MATHS
STUDENT BOOK

Craig Lowther • Judith Barron • Brenda Harden
Jennifer Smith • Alysoun Wilson

001/22042015

10 9 8 7 6 5 4 3 2 1

ISBN 9780007504633

Published by
Leckie & Leckie Ltd
An imprint of HarperCollins*Publishers*
Westerhill Road, Bishopbriggs, Glasgow, G64 2QT
T: 0844 576 8126   F: 0844 576 8131
leckieandleckie@harpercollins.co.uk   www.leckieandleckie.co.uk

Publisher: Peter Dennis
Project manager: Craig Balfour

Special thanks to
Jouve (layout)
Ink Tank (cover design)
Lucy Haddfield (proofread)
Louise Robb (proofread)
Project One Publishing Solutions (project management and copy-editing)
Peter Lindsay (answers)

Printed in Italy by Grafica Veneta S.p.A

A CIP Catalogue record for this book is available from the British Library.

Acknowledgements

This product uses map data licensed from Ordnance Survey © Crown copyright and database rights (2014) Ordnance Survey (100018598) – OS Map (Chapter 12)
Time zone map (Chapter 15) and ferry map (Chapter 22) © CollinsBartholomew
Edinburgh airport (chapter 15) © Brendan Howard / Shutterstock.com

Illustrations © HarperCollins Publishers

All other images © Shutterstock.com

ANSWERS

Answers to all Exercise questions are provided online at www.leckieandleckie.co.uk/N5Lifeskills or go to www.leckieandleckie.co.uk, click on the 'Free Resources' button and navigate to 'N5 Lifeskills Maths'.

Introduction

About this book

This book provides a resource to practise and assess your understanding of the mathematics covered for the National 5 Lifeskills Maths qualification. There is a separate chapter for each of the skills specified in the Managing Finance and Statistics, Geometry and Measure and Numeracy units. Most of the chapters use the same features to help you progress.

You will find a range of worked examples to show you how to tackle problems, and an extensive set of exercises to help you develop the whole range of operational and reasoning skills needed for your National 5 assessment.

The book also contains a number of short, medium and extended case studies, as well as a section devoted to preparation for assessment to ensure that you are fully-prepared for your assesment throughout the course.

You should not work through the book from start to finish. Your teacher will choose a range of topics throughout the school year and teach them in the order they think works best for your class, so you will use different parts of the book at different times of the year.

Features

CHAPTER TITLE

The chapter title shows the skill covered in the chapter.

12 Planning a navigation course

THIS CHAPTER WILL SHOW YOU HOW TO:

Each chapter opens with a list of topics covered in the chapter and tells you what you should be able to do once you have worked your way to the end.

This chapter will show you how to:

- use bearings to create an accurate plan
- create and interpret an accurate plan
- use bearings and the relationships between speed, distance and time to solve navigation problems.

YOU SHOULD ALREADY KNOW:

After the list of topics covered in the chapter, there is a list of topics you should already know before you start the chapter. Some of these topics will have been covered before in Maths, and others will depend on having worked through different chapters in this book.

You should already know:

- that bearings are measured from north in a clockwise direction and have three figures
- how to convert between metric units
- how to read and apply scales
- how to use measuring instruments accurately.

EXAMPLE

Each new topic is demonstrated with at least one worked Example, which shows how to tackle the questions in the Exercise that follows it. Each Example breaks the question and solution down into steps so you can see what calculations are involved, what kind of rearrangements are needed and how to work out the best way of answering the question. Most Examples include comments, which help explain the processes.

Example 12.1

Masood travels on a bearing of 075° for 300 metres.

Use a scale of 1 cm : 50 m to show Masood's journey.

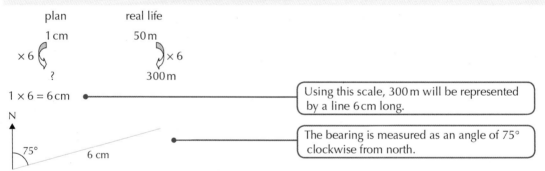

Using this scale, 300 m will be represented by a line 6 cm long.

The bearing is measured as an angle of 75° clockwise from north.

EXERCISE

The most important parts of the book are the Exercises. The questions in the Exercises are carefully graded in difficulty, so you should be developing your skills as you work through an Exercise. If you find the questions difficult, look back at the Example for ideas on what to do. Use Key questions, marked with a star ★, to assess how confident you feel about a topic. Questions which require reasoning skills are marked with a cog icon ⚙.

ANSWERS

Answers to all Exercise questions are provided online at www.leckieandleckie.co.uk/N5Lifeskills **or** go to www.leckieandleckie.co.uk, click on the 'Free Resources' button and navigate to 'N5 Lifeskills Maths'.

ACTIVITY

Activities are placed at the end of each exercise, or set of exercises. These enable individual, small-group or whole class activities to further investigate the topic's themes.

GO! Activity

The plan below shows the dimensions of a new supermarket.

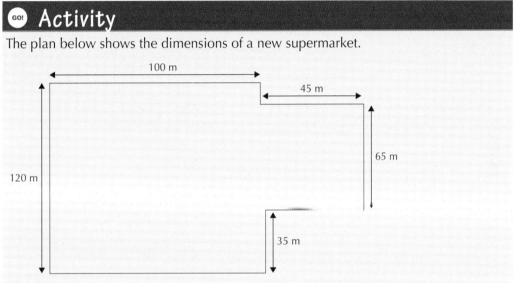

a   Decide where the entrance, exit and checkouts will be placed.

b   By visiting a local supermarket or looking on the internet, research some types of products sold by a supermarket. Investigate how much shelf space is given to each type of product.

c   Make a sketch of the design showing the position of each product.

d   Using an appropriate scale, make a scale drawing of the supermarket.

e   Why do you think some products are placed where they are? Give some reasons.

HINTS

Where appropriate, Hints are given in the text and in Examples to help give extra support.

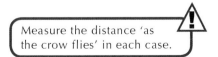

Measure the distance 'as the crow flies' in each case.

END-OF-CHAPTER SUMMARY

Each chapter closes with a summary of learning statements showing what you should be able to do when you have completed the chapter. The summary identifies the Key questions for each learning statement. You can use the End-of-chapter summary and the Key questions to check you have a good understanding of the topics covered in the chapter.

- • I can complete accurate enlargements and reductions of diagrams using a given scale factor. ★ Exercise 11A Q1

- • I can interpret scale diagrams. ★ Exercise 11B Q6, Q12

- • I can construct scale diagrams. ★ Exercise 11C Q1

- • I can explain decisions based on the results of measurements. ★ Exercise 11C Q7

ASSESSMENTS

A preparation for assessment section and case studies section are provided for each of the three units. These cover the minimum competence for the unit content and are good preparation for your unit assessment.

1 Analysing a financial position using budget information

This chapter will show you how to:

- analyse a financial position using a budget
- use a budget to plan an event
- balance incomings and outgoings.

You should already know:

- how to read and interpret simple budget information
- how to balance simple incomings and outgoings.

Budgeting and planning for personal use

There are a number of things to think about when using a budget:

- How much money do you have to start off with?
- How much money do you need?
- What do you do if you don't have enough money (known as a **deficit)?**
- What do you do if you have more money than you need (known as a **surplus**)?

The graph below shows the average weekly expenditure by UK households in 2011. This provides an estimate of essential and non-essential spending.

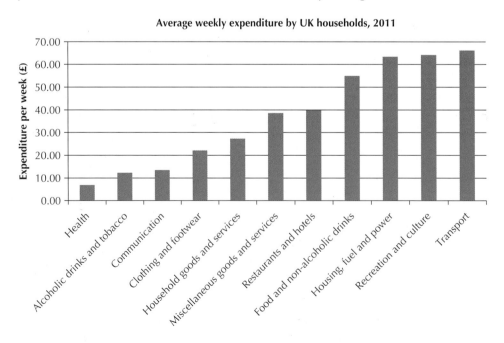

Average weekly expenditure by UK households, 2011

How much money do you have to start off with?

Consider your current **income**. Income might include pocket money or allowance, money earned by doing chores for your family and neighbours, a part-time job and savings.

Anyone aged 16 or more must pay income tax and National Insurance contributions if they earn more than £10000 in the year (this is based on the tax year 2014–15). **Net pay** (also known as take-home pay) is the amount of money received from your employer once tax or other deductions are made. This is explored in more detail in Chapter 2.

Budgeting is always based on your net pay. It is helpful to budget based on regular income. For example, if you are paid weekly (or receive pocket money or allowance weekly), then you should make a weekly budget.

For the most up-to-date information on a tax year, refer to the HMRC website: www.hmrc.gov.uk

How much money do you need?

Spending is based on essential and non-essential items. Here is an example of Fiona's monthly expenditure (outgoings).

Monthly expenditure	£
Mortgage	605
Gas/electricity	105
Council tax	98
TV/phone/internet	49
Petrol	250
Food	125
Insurance	55
Entertainment	140
Clothing	50
Savings	120
Total	**1597**

Essential outgoings include food, mortgage (or rent/family contributions), gas and electricity, council tax, petrol (or other transport costs) and insurance.

Non-essential items include TV/phone/internet, eating out, entertainment and clothing.

When budgeting, identify the difference between essential and non-essential spending. Then build savings in to the budget so that unexpected expenses can be paid for (for example, car repairs or increased outgoings), as well as luxury items (for example, holidays).

What do you do if you don't have enough money?

If you do not have enough money, refer back to your list of essential and non-essential expenses. Identify ways of reducing expenditure and prioritise your expenditure accordingly.

For example, in Fiona's budget above, she could easily reduce the amounts budgeted for clothing and entertainment, but her essential spending needs more detailed consideration. She might be able to change her energy usage in order to lower her gas/electricity costs or she could change the way she shops for food in order to reduce those costs.

If you still don't have enough money to cover your costs, you may need to take more extreme measures, such as moving home to reduce mortgage or rent payments. Many young adults choose to move back to their parents' homes (or have never moved out) due to increased living costs: it is estimated that around half of 20–34 year olds still live in their parental home as they cannot afford to rent or buy their own home (*Mirror*, 29 July 2014).

What do you do if you have more money than you need?

If you have more money than you need (surplus), you have a choice of saving or spending, or a bit of both. You could:

- save money, for example by putting it in a high interest account (Chapter 5 covers this in more detail).
- increase essential spending. For example you could increase mortgage payments as a way of reducing the overall cost of paying back a mortgage or you could move to a larger property if you can afford an increase in mortgage or rent payment.
- increase non-essential spending, including clothing, holidays and entertainment.

Always remember that it is important to include savings in any budget. If you own a car or a house, you can expect to have to pay for unexpected expenses at some stage (such as new tyres for the car or replacing a washing machine).

Example 1.1

Sam receives an Educational Maintenance Allowance by attending school in his fifth year. He receives £30 per week. His mother gives him £5 per week and he also does a 3-hour weekly paper round on a Friday evening, earning £3.80 per hour.

Sam plans this weekly budget: £20 for entertainment, £15 for clothing and £9 for his mobile phone.

a Calculate his weekly income.

b Based on this budget, how much can Sam save each week?

c He wants to buy a ticket to see his favourite band play at a gig next year. The band is very popular and he needs to have enough money saved by the date when the tickets go on sale. Tickets cost £75.

   i If tickets go on sale in 15 weeks' time, will he have enough money saved to buy a ticket? Give a reason for your answer.

   ii If he reduces his clothing budget to £10 per week, can he afford to buy the ticket?

a Income from paper round is £3.80 × 3 = £11.40

   Weekly income is £30 + £5 + £11.40 = £46.40

b Weekly expenditure is £20 + £15 + £9 = £44

   Sam saves £46.40 – £44.00 = £2.40 per week

c i 15 × £2.40 = £36

   No, Sam cannot afford to buy the ticket as he will only have saved £36, which is less than £75.

   ii New weekly expenditure is £20 + £10 + £9 = £39

   Sam now saves £7.40 per week

   15 × £7.40 = £111

   Yes, he can now afford to buy the ticket as he will have saved £111, which is more than £75.

Example 1.2

Here is Jenny's recent bank statement.

Bank of Stirling		Statement for J. Watt 2 June	
Date	**Details**	**Money in**	**Money out**
1 May	Stirling Council (Salary)	£1625	
1 May	Mortgage		£605
2 May	Petrol		£75
2 May	Supermarket		£45
4 May	Council tax		£98
6 May	ATM		£40
8 May	TV/phone/internet		£49
10 May	Petrol		£70
12 May	Insurance		£55
14 May	Clothes		£50
15 May	ATM		£30
17 May	Petrol		£50
17 May	Gas/electricity		£105
18 May	Supermarket		£48
20 May	ATM		£30
25 May	Petrol		£55
27 May	Supermarket		£32
29 May	ATM		£40

The ATM withdrawals cover her entertainment and incidental daily costs.

a If this is a typical month for Jenny, how much can she save each month?

b i Jenny wants to buy a new car. Her car loan will cost her £170 per month. Can she afford to buy the new car? Give a reason for your answer.

   ii Jenny would like to be able to save a minimum of £40 per month. Give an example of how she could reduce other expenditure in order to save £40 per month and still be able to buy the new car.

(continued)

a

Monthly expenditure	£
Mortgage	605
Gas/electricity	105
Council tax	98
TV/phone/internet	49
Petrol	250
Supermarket	125
Insurance	55
ATM	140
Clothing	50
Total	**1477**

Jenny can save £1625 − £1477 = £148 per month

b   i   No, because she can only save up to a maximum of £148 which is less than £170.

   ii   Jenny could make these changes to her spending to allow her to save £40 per month and still afford the new car.

Monthly expenditure	£
Mortgage	605
Car loan	170
Gas/electricity	105
Council tax	98
TV/phone/internet	49
Petrol	250
Supermarket	113
Insurance	55
ATM	110
Clothing	30
Savings	40
Total	**1625**

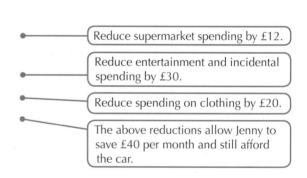

Reduce supermarket spending by £12.

Reduce entertainment and incidental spending by £30.

Reduce spending on clothing by £20.

The above reductions allow Jenny to save £40 per month and still afford the car.

Exercise 1A

★ 1   John is going to the cinema with friends. His mother gives him £15 to spend. The cinema ticket costs £6.30. Before going to the cinema he has a burger costing £2.89 and a bottle of juice costing £1.25. He also buys a football magazine for £1.89. His bus fare costs £1.80 return.

   a   How much money has he left to spend on popcorn at the cinema counter?

   b   If he wants to buy a box of popcorn worth £2.99, how much money will he need to borrow from a friend?

   c   How could John have spent his money differently to enable him to afford to buy the popcorn (and avoid borrowing money from his friend)?

2   Azam is in S6. He makes a note of his weekly income and expenditure.

Weekly income	Weekly spending
£10 weekly allowance from parents	£15 entertainment with friends
Waiter Friday and Saturday evenings (4 hours each evening), paid £3.80 per hour	£10 clothing
	£8 mobile phone

   a   Calculate Azam's weekly income.

   b   Based on this budget, how much can Azam save each week?

   c   He wants to buy a tablet computer costing £230. If he puts all of his savings towards buying this computer, how many weeks will it take him to be able to buy it?

   d   If he decides to reduce his expenditure to spending £12 per week on entertainment and £5 per week on clothing, how much sooner will he be able to buy the computer?

3   Sean works as a childcare assistant in a nursery and earns £840 per month.
   The table shows his outgoings.

Monthly expenditure	£
Car loan	160
Petrol	240
Parking	120
Car repairs/service	30
Contribution to family expenses	150
Car insurance	45
Entertainment	70
Clothing	20

   a   Calculate how much Sean can save each month.

   b   Sean is saving money because he would like to move out of his parents' house and buy his own flat. At the moment he is budgeting only a small amount per month as savings. He notices that a lot of his monthly wage is spent on his car.

      i   How much does he spend on his car?

      ii   If he gives up his car and buys a monthly travel pass for £150, how much can he now save?

   c   Sean does give up his car and buys the monthly travel pass.

      i   If he increases his entertainment spending to £80 and his clothing budget to £50, how much can he now save per month?

      ii   How long will it take him to save £5000 towards a deposit on a house?

4   Frances works as an apprentice car mechanic and currently earns £720 per month.
    The table shows her monthly expenditure.

Monthly expenditure	£
Car loan	120
Petrol	150
Car repairs/service	40
Contribution to family expenses	?
Car insurance	65
Entertainment	100
Clothing	40
Savings	25
Total	**720**

a   How much does she contribute to the family expenses?

b   Frances has been offered a room in a shared flat with friends. She makes a list of
    changes to her expenditure.

Changes to expenditure

Share of rent £250

Share of gas/electricity £45

Share of council tax £60

Monthly salary increase to £890 (final year of
    apprenticeship)

Petrol £120 (living closer to work)

Car insurance increase to £75 (change of address)

Food £120

> ⚠ Car insurance costs
> depend on where
> you live, as some
> areas are considered
> as higher risk than
> others due to police
> crime statistics.
> Another factor is
> whether the car is
> parked in a garage, on
> a private driveway or
> is parked on the street.

i   Copy and complete the following table for Frances's
    estimated expenditure, assuming her other expenditure remains the same.

Monthly expenditure	£
Car loan	
Petrol	
Car repairs/service	
Rent	
Gas/electricity	
Council tax	
Car Insurance	
Entertainment	
Clothing	
Food	
Savings	

ii   Can she afford to move out of home? Give a reason for your answer.

iii   Name two things that she could do in order to be able to reduce her expenditure.

c   If she sold her car and bought a travel pass for £130, would she now be able to afford to move out of home? Give a reason for your answer.

5   Duncan is a student and works in a bar. He usually works 50 hours in a month, earning £4.20 per hour. He also receives an annual student support package of £1800 and takes out a student loan of £4500 for a year. Duncan also receives £60 per month from his parents. In order to draw up a budget he calculates his monthly income and expenditure.

⚠ Divide the annual student support package and student loan contributions by 12.

a   Calculate Duncan's monthly income.

b   Duncan lives in a shared flat. His monthly expenses are listed below.

Monthly expenditure	£
Rent	240
Gas/electricity	43
Council tax	55
Food	130
Clothing	50
Entertainment	120
Books	25
Travel pass	120

Calculate Duncan's monthly savings.

c   Duncan wants to go on holiday with a group of friends once his exams are over. The holiday is all-inclusive and will cost him £625.

i   If he extends his hours in the bar by an extra 10 hours each month, how much can he now save each month?

ii   If he saves this amount each month for 6 months, will he save enough to go on holiday?

iii   What other holiday options do Duncan and his friends have? Research the prices of other options in holiday brochures or on the internet.

6   Joe has a couple of part-time jobs.

He does a weekly paper round and gets paid £18 per week.

He also works on a market stall from 8 a.m. until 2 p.m. every Saturday and gets paid £3.80 per hour.

He enjoys going out with his friends and would normally spend about £15 per week on entertainment.

He pays for his mobile phone which costs him £10 per week and buys magazines which cost him around £5 per week.

He saves the rest of the money he earns.

a   Calculate Joe's weekly income.

b   Calculate his weekly expenditure.

c   Joe wants to buy a bicycle costing £239.99. How many weeks would he need to save in order to purchase the bicycle?

★ 7 Joanne and Billy are married and have two children. Joanne earns £1165 per month and Billy earns £1546 per month. They also receive child benefit and family tax credits totalling £186.25 per month.

a Calculate their monthly income.

b The table shows the family's expenditure each month.

Monthly expenditure	£
Mortgage	640
Gas/electricity	145
Council tax	155
TV/phone/internet	75
Petrol	240
Food	525
Car loan	242
Car insurance	68
Entertainment	150
House and contents insurance	54
Clothing	100
Children's pocket money	50
Mobile phones	45

Using the figures in the table, how much can they save each month?

c Joanne and Billy currently have savings of £2246.

They would like to save for a new kitchen and avoid taking out a loan.

The kitchen they want to buy costs £4400 and it will cost £2450 to have it fitted.

How long will they have to save before they can get a new kitchen?

8 Sylvia is in S5 at school. She makes a list of her weekly income and expenditure.

Weekly income	Weekly spending
£20 per week bursary	£5 mobile phone contract
Supermarket job: 2 hours Tuesday evening, 8 hours Saturday, earning £4.10 per hour	£12 gym membership
	£2.35 lunch for every school day
	£8 clothing (second-hand shops and internet shopping)
	£15 entertainment with friends

a Calculate:

i Sylvia's weekly income

ii Sylvia's weekly expenditure

iii the amount she can save each week.

b What can she do to reduce her weekly outgoings and increase her savings?

9 Sid is on a Jobseeker's Allowance which amounts to £71.70 per week. He receives housing benefit, therefore does not pay rent.

He makes a note of his weekly expenditure.

Weekly expenditure	
Council tax	£2.21
Gas/electricity	£18
Food	£30
Pay-as-you-go mobile	£5
TV licence	£4

> Council tax is normally £26, but Sid gets a 91.5% reduction because he is on jobseeker's allowance.

a   By calculating his income and expenditure, how much does Sid have left to spend on clothing and entertainment and for savings?

b   He wants to buy a new suit for a job interview. The suit costs £120. If he spends £6 per week on entertainment, how long will it take him to save up for his new suit?

c   What other support is available to Sid to help him buy a new suit?

10   Here is William's recent bank statement. He works for an insurance company but he also runs the local pub quiz one night a week (paid monthly).

Leckie Bank		Statement for W. Wilson 2 September	
Date	Details	Money in	Money out
1 Aug	Friends Inc. (salary)	£1725	
1 Aug	Rent		£655
2 Aug	Petrol		£65
2 Aug	Supermarket		£40
4 Aug	Council tax		£104
4 Aug	ATM		£50
5 Aug	TV/phone/internet		£53
8 Aug	Petrol		£65
10 Aug	Insurance		£68
10 Aug	Clothes		£75
12 Aug	ATM		£40
15 Aug	Quizmaster	£80	
17 Aug	Petrol		£65
18 Aug	Gas/electricity		£125
20 Aug	Supermarket		£42
21 Aug	ATM		£40
23 Aug	Petrol		£55
26 Aug	Supermarket		£37
27 Aug	ATM		£50

a   This is a typical month for William. Does he make a surplus or deficit in the month and by how much?

b   William's rent increases by 5% and his gas/electricity increase by 12%. He also loses his job as quizmaster. Does he now make a surplus or deficit in the month and by how much?

c   What can William do to increase his savings and reduce his monthly outgoings?

11 Alex is a student and has already spent her student loan and annual student support package issued the previous September. She has a part-time job and receives a monthly allowance from her parents.

Collins Bank		Statement for A. Long 3 April	
Date	Details	Money in	Money out
1 Mar	Mr J. Long (parent)	£100	
1 Mar	Rent		£220
3 Mar	Supermarket		£35
4 Mar	Council tax		£46
4 Mar	ATM		£50
6 Mar	Phone/internet		£10
7 Mar	ATM		£20
9 Mar	Clothes		£25
10 Mar	ATM		£40
12 Mar	Burgers R Us (wage)	£242	
15 Mar	Travel pass		£49
15 Mar	Gas/electricity		£40
19 Mar	Supermarket		£38
21 Mar	ATM		£50
23 Mar	Bank charge		£10
25 Mar	Supermarket		£27
29 Mar	ATM		£40

At the start of the month Alex's account is £100 overdrawn. The bank charges Alex £10 for being overdrawn.

a   Does Alex make a surplus or deficit this month and by how much?

b   When Alex is £1000 overdrawn the bank refuses to allow her to withdraw any money. If Alex spends a similar amount in future months, how long will it take her to reach an overdraft of £1000?

c   What could Alex do to reduce her expenditure?

d   What could Alex do to increase her income?

12 Jacob is 15 years old and lives in Burkina Faso (West Africa). In Burkina Faso the currency is Central African Francs (fr).

Here is a list of his income and expenditure for one month.

Income	Expenditure
Sells grain for his father on Saturdays and Sundays	School fees 5000fr
1 kg bag of grain costs 600fr	Medical costs 12000fr
Sells 21 bags this month	School book 2800fr
	Pair of shoes 3000fr

a   By calculating Jacob's income and expenditure, does he have enough money to pay for his expenses this month? Give a reason for your answer.

b   How many more bags will he need to sell in order to pay for all his expenses?

c   What else could Jacob do to reduce his expenses, increase his income or pay his expenses using an alternative method?

Using a budget to plan an event

When planning an event, you should ask the same questions as you did when thinking about your personal budget:

- How much money do you have to start off with?
- How much money do you need?
- What do you do if you don't have enough money?
- What do you do if you have more money than you need?

How much money do you have to start off with?

You might not have any money to start with, in which case you need to plan effectively to ensure that any money you need to pay has been generated by the time you need to pay it. Depending on the nature of the event, you may find that the money needed will come from ticket sales or the cost per head based on budgeting calculations.

For instance, if you are planning to go on holiday you need to ensure that you can pay for the holiday by the time it needs to be booked and paid for. Make sure you budget effectively in order to ensure that you have the necessary funds, and always prepare for any unexpected spending.

How much money do you need?

Ensure that you plan carefully, take into account all planned spending, and always allow for some extra spending.

For example, if you are organising a social event and planning the budget for it, make sure that your cost per ticket is based on fewer people than you expect to attend in case you have any unexpected costs (or in case you don't sell as many tickets as you hoped).

What do you do if you don't have enough money?

Refer back to your original budget and identify ways of reducing expenditure. Prioritise your expenditure (list your spending in order of importance), so you make sure the most important things are budgeted for. This is especially important if you are planning an event on behalf of others: if the event makes a loss, this may affect others and you will be asked to explain why the loss occurred.

For example, if you are planning a birthday party and you realise that your planned spending is more than you budgeted for, you may need to:

- change the venue
- reduce the amount spent per head on food and drinks
- spend less on entertainment
- spend less on decorations

What do you do if you have more money than you need?

If you are planning for an event which is for yourself, this is straightforward: you have budgeted for more than you need therefore you can put extra money into savings or you can spend more on the event itself.

However, if you are planning an event for a school or organisation, this means that the surplus can be returned to individuals or the money can be added to funds for future events or given to charity if appropriate.

Example 1.3

Isla is planning a holiday with friends. She can afford to save £60 each month for 6 months. She has asked her parents for money for her birthday which she will put towards her holiday. Her parents say they will give her £200. At Christmas Isla receives £150 from family members. She also has £54.20 in her bank account that can be put towards the holiday.

a   How much money can she spend on the holiday?

The cost of the holiday is £495 per person for 7 nights in Turkey which covers the cost of the accommodation and flights plus half-board (dinner, bed and breakfast but not lunch).

Isla will need spending money for entertainment, lunches and souvenirs. She plans to put aside £200 for this (which she will exchange into local currency).

Holiday insurance costs £15.45.

Isla has written a list of the clothes and toiletries she wants to buy for the holiday.

Clothes	Toiletries
• Swimsuit £18	• suntan lotion, £8.99
• 3 pairs of shorts, £8.99 each	• after-sun lotion, £6.99
• 4 Suntops - twinpacks cost £12	• insect repellent, £4.65
• 2 dresses, one for £18, the other for £9.99	

b   Calculate Isla's estimated expenditure. Can she afford the holiday? Give a reason for your answer.

c   Suggest how Isla could reduce expenditure in order to be able to afford the holiday.

a

Income	£
Savings = £60 × 6	360.00
Birthday	200.00
Christmas	150.00
Bank savings	54.20
Total	**764.20**

b

Expenditure	£
Flights and accommodation	495.00
Spending money	200.00
Insurance	15.45
Toiletries	20.63
Swimsuit	18.00
Shorts	26.97
Suntops	24.00
Dress	27.99
Total	**828.04**

No, she cannot afford the holiday based on this expenditure as the cost of the holiday is £828.04 which is greater than the amount Isla has budgeted for (£764.20).

c

Expenditure	£
Flights and accommodation	495.00
Spending money	160.00
Insurance	15.45
Toiletries	20.63
Swimsuit	18.00
Shorts (2 pairs)	17.98
Suntops	24.00
Dresses	9.99
Total	**761.05**

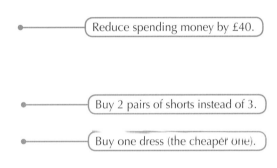

Reduce spending money by £40.

Buy 2 pairs of shorts instead of 3.

Buy one dress (the cheaper one).

She could also:

- shop around for cheaper holiday insurance
- share toiletries with friends to reduce costs
- reduce clothes spending a different way, such as buy fewer suntops.

Exercise 1B

1   A youth group is organising their disco. The group needs to ensure that it breaks even or makes a profit. They make a list of possible expenditure for the evening.

> **Disco expenditure**
>
> DJ: £120 (brings own equipment)
>
> Hall: £75 for 3 hours
>
> Crisps: sold in boxes of 48, costing £20 per box
>
> Bars of chocolate: sold in boxes of 48, costing £30 per box
>
> Cans of juice: sold in packs of 24, costing £10 per pack

A packet of crisps, a bar of chocolate and a can of juice is included in the price of the ticket.

a What is the minimum ticket price the group should charge in order to break even if it expects:

   i 200 people

   ii 250 people

   iii 300 people?

Round all your answers to the nearest 10p.

b The group wants to make a profit of £500 to give to charity. What is the minimum ticket price it should charge in order to break even if it expects:

   i 200 people

   ii 250 people

   iii 300 people?

Round your answers to the nearest 10p.

2 The Enterprise group in a school decides to make clocks out of CDs. It needs to:
- purchase clock mechanisms at £2.99 each
- purchase CDs costing £25 for 100.

The group will sell the clocks for £4.50 each.

a If it makes 200 clocks and sells them all, how much profit will the group make?

b If it makes 200 and sells only 140, will the group make a profit?
Give a reason for your answer.

c If it does not sell all of the clocks made, what recommendation would you give the group in order to try to break even?

3 A teacher is planning a school trip to an outdoor centre.
- The trip is planned for 40 S2 pupils and 4 teachers.
- The outdoor centre charges £26 per person per night.
- The trip is planned for 3 nights.
- The bus to transport all pupils and staff costs £190 each way.
- In addition, the school needs to pay £50 to use the inflatable 'Bungee Run' one afternoon.
- Four teachers need to attend the trip and the cost of their trip is covered by the pupils' payments.

How much does each pupil need to pay in order to cover all costs?

Your answer should be expressed as a whole number of pounds.

4 The school prom committee is organising this year's prom at a local hotel for 75 pupils and 15 staff members.

The committee makes this list of its expenditure for the evening.

Prom expenses	£
Meal	£24 per person
Band	790.00
Photo booth	450.00
Balloons	8.45
Helium canister	18.99
Other decorations (confetti, party poppers, streamers)	14.26

The staff who attend the prom will pay £15; the remainder of their expenses must be covered by the price of the pupils' tickets.

How much should the committee charge each pupil to ensure that the committee does not make a loss? Express the ticket price as a whole number of pounds (£).

★ 5 Josh is planning to go to a weekend music festival in Scotland with his friends. The table shows his budget.

Music festival expenditure	£
Ticket (covers Friday, Saturday, Sunday)	193.95
Camping @ £4.50 per night for 4 nights (Thurs–Sun)	
Food and drink: £20 per day for 4 days (Thurs–Sun)	
Breakfast on Monday	5.00
Bus journey: £5.50 each way	
Tent	
Sleeping bag	12.50
Total	

Josh and his friend Billy agree to buy a two-man tent for £46. They will each pay half of the cost.

a Copy and complete the table of expenditure. How much will it cost Josh to go to the music festival?

b Josh decides to earn money in order to pay for the festival. He will wash cars and mow lawns.

- He charges £5 per carwash and £10 for each lawn he mows.
- He washes 25 cars and mows 16 lawns.

In addition, his parents give him £30, his grandmother gives him £15 and his uncle gives him £10.

Does he have enough money to go to the festival? Give a reason for your answer.

6 A youth club organises a 5-a-side football tournament for other youth clubs to participate in.

- 15 teams intend to participate.
- Each team has five players and one reserve.
- They hire five pitches. Each pitch costs £18 an hour to hire and the tournament lasts for 3 hours.
- Each player is given a water bottle and healthy snack.

- Water bottles are bought in packs of six for £1.50. Snack bars are bought in packs of five for £1.20.
- Medals are bought for the winning team at a cost of £1.20 per medal. Each member of the winning team (including the reserve) wins a medal.
- The winning team is also given a £50 voucher for sports equipment for their youth club.

How much should each team member be charged so that the costs are covered?

Your answer should be expressed as a whole number of pounds.

★ 7 A school is putting on a musical in which pupils can participate. The musical will take place for three nights.

a Based on previous school musicals, the school expects to sell 120 adult tickets and 80 concession (under 16) tickets for each performance.

They intend to sell the tickets at a cost of £8 per adult and £5 per concession.

Calculate their expected income.

The expenditure is shown in the table.

School musical expenditure	£
Use of school hall	75
Props	165
Scenery	265
Costumes: £12 per pupil (46 pupils)	
Sound equipment	260
Lighting	350
Royalties: 15% of ticket sales	
Printing tickets	28
Total	

'Royalties' are the payment to the composer and author for allowing the use of their material.

b Copy and complete the table. What is the school's total expenditure?

c Calculate their expected profit.

8 Amara is planning her 18th birthday party. She makes a list of her expenses.

Birthday party expenses	
Hall rental	£125
Buffet for 120 people	£4.95 per person
DJ	£225
Cake	£25
Balloons	£9.65
Helium canister	£21.45
Other decorations	£14.28
(confetti, party poppers, streamers)	

Amara and her father have agreed to split the cost of the party between them in the ratio of 3 : 2

a How much should Amara expect to pay?

b In order to afford the cost of the party, Amara has asked her friends to give her money rather than presents.

Of the 120 people invited to the party, 35 are family members and the remainder are her friends. Approximately how much will Amara need to get from each friend so that she can cover the cost of the party?

9   A school holds a summer fun day. As part of the fun day, they run a fast food stall selling burgers in rolls.

They make a list of the food needed and they check packaging and prices with the local cash and carry. They also record how many of each they expect to sell based on last year's sales.

Name of food	Sold as	Cost	Expect to sell
Meat burger	pack of 8	£1.15 per pack	350
Veggie burger	pack of 6	£1.58 per pack	50
Roll	pack of 6	65p per pack	400
Margarine	1 kg tub	£2.80 per tub	3 tubs
Tomato ketchup	700 g bottle	£1.80 per bottle	5 bottles

a   Calculate the total expenditure for the burger stall.

b   The meat burgers and veggie burgers are sold at the same price. How much will they need to charge per burger in order to:

   i   break even

   ii   make a minimum profit of £100?

● Activity

1   For the next month record your own income and expenditure and put the information into a table, similar to the tables used in Exercise 1A. How much are you saving? Is there a way you can reduce your spending if there is something you want to save for?

2   You are going to plan a 6-month budget. To do this you need to draw up a monthly budget and look at saving and spending over the 6 months.

You have decided to save at least 20% of your income each month.

Here are the things that you can spend your money on.

Chocolate bar 60p	**Music download** £9.99 per album	**DVD** £7
Clothes: **Jeans** £39.99 **T-Shirt** £9.99	**Bag:** Handbag £10 or Man-bag £10	**Tablet computer** £139
Cinema ticket £6 **Popcorn/snack** £5.50	**Large bag of crisps** £1.29	**Bottle of juice** £1.15
Bicycle £90	**Pizza** Large £11.99 Medium £8.99 Small £5.99	**Burger, fries and drink** £4.25

GO! Activity

Month 1 Income: Allowance: £5 per week Birthday money: £30 Other money: £4.50	**Month 2** Income: Allowance: £5 per week
Month 3 Income: Allowance: £5 per week Other money: £8	**Month 4** Income: Allowance: £6 per week (increased)
Month 5 Income: Allowance: £6 per week Christmas money: £40 Other money: £7.45	**Month 6** Income: Allowance: £6 per week

£5 × 4 = £_____ for the month

a   Make an income and expenditure table based on the information given. You can only spend money on items on the list above. Use your monthly expenditure in the first activity to help you think about the way you spend your money per month.

b   Write down the monthly balance of your savings account.

c   At the end of the 6 months, you have to spend £40 on your grandfather's birthday present. After deducting this from your savings, how much money do you have left? What else would you choose to buy using your savings? How much money would you keep in your savings?

d   Once you have completed this task, share your responses with those of others in your class. Are there similarities? Are there differences? Would you change your monthly expenditure now that you have compared it with others?

Make the Link

Support for low income earners

Exercise 1A Q9c mentions other support available to those on a low income. It can be hard to budget on a low income, and it can be challenging to plan for a necessity such as a suit to wear to an interview.

The Citizens Advice Bureau and other support agencies can help people budget more effectively and make the most of their money. Charities help with food parcels, for example, foodbank projects. Second-hand clothing can be bought at charity shops. You can buy nearly new clothes for a fraction of the cost of new items on websites such as Ebay or Gumtree.

Many people use credit unions. A credit union is a small, non-profit organisation set up by members with something in common, for example, living in the same community or working in the same industry. A credit union does not owe profit to shareholders, so it can offer help to those who can't access ordinary bank products. It tends to offer savings and loans, but some now also offer current accounts.

Make the Link

Developing countries

Exercise 1A Q12 describes the situation of Jacob in Burkina Faso. The economy in many countries works differently from the economy of richer countries, such as Scotland. There are a couple of options available to Jacob.

- He could **barter**. Bartering means exchanging goods or services without money changing hands. So Jacob could offer items (for instance, his grain) instead of money to pay for necessities. Some schools accept grain or animals as payment for fees instead of money.

- He could **haggle**. Haggling is another word for negotiation and means that the buyer and seller can argue the price of goods or services and come to an agreement on price. For example, Jacob could start with a higher price per bag of grain, and then, after haggling with a customer, agree to a final price that they are both happy with (but which is higher than the original price of 600fr per bag). Similarly, he could haggle the price of the shoes he wants to buy from the seller.

- I can analyse a financial position using a budget.
  ★ Exercise 1A Q1

- I can use a budget to plan an event.
  ★ Exercise 1B Q5

- I can balance complicated incomings and outgoings.
  ★ Exercise 1B Q7

For further assessment opportunities, see the Case Studies for Unit 1 on pages 136–140.

2 Analysing and interpreting factors affecting income

Income

In order to analyse and interpret factors affecting income, you need to understand the meaning of different terms.

Income is money earned through employment or unearned income through interest on savings, investment, shares, pension, inheritance and public welfare payment.

Basic pay is the pay received before taking into account any additional benefits or bonuses (including overtime), and before deductions (such as tax).

Gross pay is the pay received including any additional benefits, bonuses and overtime, but before deductions (such as tax).

Net pay is the gross pay minus deductions.

A **bonus** is an extra amount added to a monthly salary, either for exceptional performance in work tasks or an extra seasonal payment, such as a Christmas bonus.

Commission is an amount added to the monthly salary of an employee usually in the sales industry and is a percentage of their monthly sales; it is used as an incentive to sell more.

Piecework is an amount paid for each item or task completed.

Benefits and allowances are regular payments from the government to help in times of sickness, disability, old age or unemployment, as well as extra payments for those who have children or who are on a low income.

A **minimum wage** is the lowest hourly, daily or monthly rate of pay that employers can legally pay employees. The national minimum wage rate is split into four categories: apprentices, 16–17 years, 18–20 years, and 21 and over.

Exercise 2A provides examples of different circumstances and career choices and their associated income.

Example 2.1

Jess works in a fashion clothes shop. She earns £6.80 per hour and works a 35-hour week.

a What is her basic pay?

b In one particular week she worked 4 hours' overtime on a Sunday at time and a half. She also received commission of 1.5% of all sales she made that week. She made £4300 of sales. What was her gross pay that week?

a Basic pay is $6.80 \times 35 = £238$

b Overtime is $4 \times 1.5 \times 6.80 = £40.80$

> 'Time and a half' means the hourly rate is 1.5 times the usual hourly rate.

Commission is 1.5% of 4300

$$= 0.015 \times 4300$$

$$= £64.50$$

Gross pay = basic pay + overtime + commission

$$= £238 + £40.80 + £64.50$$

$$= £343.30$$

Exercise 2A

★ 1 Sandra is a social worker earning an annual salary of £31 500 and Bill is an administrator for an IT company earning £25 650 per year.

a i What is Sandra's monthly gross pay?

   ii What is Bill's monthly gross pay?

b Sandra and Bill have two children, Zac (aged 8), and Tom (aged 4). They receive child benefit from the government. Child benefit is paid as follows:

   • £20.30 per week for the eldest child

   • £13.40 per week for every other child.

   i How much do Sandra and Bill receive in child benefit in a month? (Assume one month = four weeks.)

   ii What is their total gross monthly household income?

2 Shona is a self-employed beautician. She charges the rates shown.

In one month she performed 7 massages, 12 manicures, 8 pedicures and 6 facials. Three customers also gave her tips of £5, £3 and £10.

a What was her gross pay for that month?

b If she earned the same amount each month, how much would she expect her gross pay to be in a year?

Shona's beauty salon
Massage £28
Manicure £16
Pedicure £21
Facial £26

3   Jamie is unemployed and is on a Jobseeker's Allowance which gives him £71.70 per week.

   a   Calculate his annual income.

      Jamie finds a job working at a supermarket. He earns £6.75 per hour and works a 37-hour week. He also receives an annual bonus of £75 in December as a Christmas bonus.

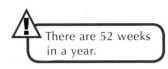

There are 52 weeks in a year.

   b   How much does he now earn per year (annual gross pay)?

   c   Compare your answers in **a** and **b**. How much more does he earn in a year when employed compared with being unemployed?

4   Katie works as a joiner. She earns £11.48 per hour and works a 35-hour week.

   She is paid overtime at the rates shown in the table.

   She recorded the overtime completed in a particular week.

Day	Overtime rate
weekday	Time and a quarter
Saturday	Time and a half
Sunday	Double time

   a   What is Katie's gross wage in this particular week?

   b   Katie has two sons. Child benefit is paid as follows:

Day	Amount of overtime
Tuesday	2 hours
Thursday	2 hours
Saturday	3 hours
Sunday	4 hours

      • £20.30 a week for the eldest child

      • £13.40 a week for every other child.

   What is her gross income in this particular week?

5   Aisha works for a company selling mobile phones and associated contracts.

   • Her basic annual salary is £18 256.

   • She earns commission of 3% of all her sales each month which is added to her basic salary.

   • Her sales in one particular month came to £31 215.

   • She also earned a bonus this month as her sales were the highest in Scotland last month. Her bonus is 0.5% of last month's sales of £34 214.

   What is her gross pay this month?

6   Iain has two jobs. He works as an art teacher two days per week and he is also an artist.

   A full-time teacher earns £34 200 per annum. Iain earns $\frac{2}{5}$ of this each year.

   a   Calculate Iain's gross monthly salary from his teaching job.

   b   Iain had paintings in an exhibition this month and sold four paintings for £1200, £950, £875 and £1340.

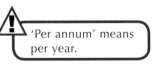

'Per annum' means per year.

      i   How much did he earn this month (before tax)?

      ii   Why would this not reflect his regular monthly income?

7 Graeme is 17 and works in a fast food restaurant. His hourly rate is £4.26 per hour.

His timesheet for last week is shown.

Day	In	Out	No. of hours worked
Monday	0700	1530	
Tuesday	0830	1715	
Wednesday	1200	1930	
Thursday	1430	2215	
Friday	1600	0000	
Saturday	1400	1830	
Total			hours

a Copy and complete the timesheet. How many hours did he work last week?

Graeme works a basic week of 37 hours.

If he works more than 37 hours, he is paid overtime at time and a quarter.

He earned a bonus of £32.50 last week as he was recommended as the friendliest server by customers in a survey.

b Graeme is also eligible for a personal independence payment (PIP) of £21.55 per week.

What was his total gross income last week?

8 Sara is a car sales executive.

- She receives an annual basic salary of £15 426.

- She also earns commission of 6% of all sales.

- In one particular month her sales total was £40 355.

a i How much was her gross pay in that month?

ii What percentage of her total gross pay was made up of his commission?

b The next month her gross pay was £4424.34. What was her sales total?

> The personal independence payment (PIP) is a tax-free payment from the government. It is a payment to assist with long-term costs as a result of long-term ill health or disability for anyone aged 16–64. The PIP replaced the disability living allowance in April 2013.

9 The chief executive of a major bank earns a monthly salary of £112 500. Last year he earned a one-off bonus of £3.375 million and performance-based bonus of £6.75 million.

He claims to work a 60-hour week.

a What was his gross annual pay last year?

b How much did he earn per hour, correct to the nearest £10?

10 Heather is a surgeon earning £117 050 per annum.

Her husband Euan works part-time as a nurse, earning £12 246 per annum.

They have four children so they receive child benefit, which is paid as follows:

- £87.96 a month for the eldest child

- £58.06 a month for every other child.

a What is their total gross monthly income?

b What percentage of their total gross income does Euan earn?

11 Iona delivers newspapers for a local newsagent. The table shows the rate she is paid.

Day	Rate per newspaper
weekday	14p
Saturday	20p
Sunday	22p

She is also allowed to keep any tips given to her by customers.

She has regular customers:

- 12 houses buy papers Monday through to Friday
- 8 houses buy papers on a Saturday
- 9 houses buy papers on a Sunday.

a In a particular week she delivers papers and received tips from three houses: £6, £3 and £2.50. What is her income in this week?

b Iona is offered a £20 bonus for every new customer provided she continues to deliver papers to them for the next 4 weeks. She gains two extra customers:

- the first customer asks for papers to be delivered from Monday through to Saturday
- the second customer asks for papers to be delivered on Saturday and Sunday.

Over the next four weeks Iona gets £42 in tips.

What is her income over these four weeks?

12 Hector is 20 and works as a cleaner. He is paid the minimum wage for his age.

Age (years)	21 and over	18–20	Under 18	Apprentice
Minimum wage (£/h)	6.31	5.03	3.73	2.68

a If Hector works a 37-hour week, what is his weekly income?

b Hector has a birthday and is now 21.

He has also been told he is now eligible for a PIP of £32.45 per week due to long-term ill-health.

   i What is his new weekly income?

   ii What is the percentage increase in his income?

⁘ Make the Link

Self-employment

In Exercise 2A Q2, Shona is self-employed. Anyone who is self-employed has to declare themselves as such to the government and then submit an annual tax return to a government agency known as Her Majesty's Revenue and Customs (HMRC). HMRC then works out how much tax this person must pay to the government. Anyone who is self-employed but does not declare their earnings to the government is breaking the law and is subject to a fine if caught.

Employment vs unemployment

In Q3, there are several factors for Jamie to consider. Now that he is working full-time he will no longer be entitled to benefits, for example, housing benefit. However, it is in Jamie's long-term best interests to gain and remain in employment so that he always has a regular income. It is generally easier to get a new job if you are already employed and being employed helps to protect you from the effects of future government cuts to the benefit system.

Deductions

This chapter explores three main types of deduction made to earned income:

- income tax
- National Insurance
- pension contributions.

Income tax

Income tax is a tax paid to the government based on income. It is deducted from the employee's wages/salary by their employer.

Not all income is taxed: you are allowed to earn a certain amount before you pay tax. This is called your **tax allowance** and is set by the government each year.

Tax allowance consists of a personal tax allowance plus blind person's allowance if eligible. Other allowances have previously been added to the tax allowance (such as disability living allowance and married person's allowance) but these are no longer accepted.

Taxable income is your annual salary minus the tax allowance.

In the year 2014–15, the personal tax allowance is £10 000 and the blind person's allowance is £2230.

In 2014–15 the basic rate of tax is 20% of taxable income up to £31 865.

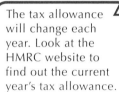

The tax allowance will change each year. Look at the HMRC website to find out the current year's tax allowance.

There is a higher tax rate of 40% for taxable income between £31 865 and £150 000, and there is an additional rate of 45% for taxable income earned over £150 000.

People who are registered as self-employed also pay income tax. They have to fill in an annual tax return and submit to the government declaring all income received.

The 2014–15 rates are used for the Examples and Exercise questions in this chapter.

Example 2.2

Lawrence is a fire-fighter. He earns a total annual income of £21 157.

His personal tax allowance is £10 000.

He pays tax at the basic rate of 20%.

Calculate:

a his taxable income

b how much income tax he should pay each year

c how much income tax he should pay each month.

a Taxable income is £21 157 − £10 000 = £11 157

b Annual income tax is 20% of £11 157 = £2231.40

c Monthly income tax is 2231.40 ÷ 12 = £185.95

Remember to round any questions involving money to 2 d.p.

Example 2.3

Julie is the headteacher of a primary school. She earns a total annual income of £52 542

Her personal tax allowance is £10 000 and she is also entitled to a blind person's allowance of £2230.

She pays tax at the basic rate of 20% for the first £31 865 of taxable income and pays the higher rate of 40% on taxable income over £31 865.

Calculate:

a   her taxable income

b   how much income tax she should pay each year.

a   Tax allowance is £10 000 + £2230 = £12 230

   Taxable income is £52 542 − £12 230 = £40 312

b   £40 312 − £31 865 = £8447

> The basic rate of 20% is paid up to £31 865, and any taxable income over £31 865 is charged at 40%. Find the amount which is taxed at 40%. The diagram represents this information.

20% of £31 865 = £6373

40% of £8447 = £3378.80

Income tax is £6373 + 3378.80 = £9751.80

Exercise 2B

1   Henry is an air traffic controller. He earns a total annual income of £29 443.
   His personal allowance is £10 000 and he pays tax at the basic rate of 20%.
   Calculate:
   a   his taxable income
   b   how much income tax he should pay each year.

2   Anya is a midwife. She earns a total annual income of £21 388.
   Her personal allowance is £10 000. She pays tax at the basic rate of 20%.
   Calculate:
   a   her taxable income
   b   how much income tax she should pay each year.

3   Sean is a paramedic. He wants to work out how much money he will have after he has paid income tax for the year. He writes these details:
   Annual income = £27 625
   Personal allowance = £10 000
   Tax paid at 20%

Calculate:

a Sean's taxable income

b how much income tax he should pay each year

c how much money Sean earns after he has paid tax.

4 Zara is a police officer. She earns a total annual income of £40 020. Her personal allowance is £10 000 and she is also entitled to blind person's allowance of £2230. She pays tax at the basic rate of 20%.

Calculate:

a her taxable income

b how much income tax she should pay each year.

5 Laura is a junior doctor and earns a total annual income of £22 400. Her personal allowance is £10 000 and she pays tax at the basic rate of 20%.

Calculate:

a her taxable income

b how much income tax she should pay each year

c how much income tax she should pay each month.

6 Sally is a lecturer at a further education college and is registered blind. She earns a total annual income of £34 250. Her personal allowance is £10 000 and she is also entitled to a blind person's allowance of £2230. She pays tax at the basic rate of 20%.

Calculate:

a her taxable income

b how much income tax she should pay each year.

c how much income tax she should pay each month.

7 Jim is a civil engineer.

- He earns a total annual income of £48 588.
- His personal allowance is £10 000.
- He pays tax at the basic rate of 20% for the first £31 865 of taxable income.
- He pays the higher rate of 40% on taxable income over £31 865.

Calculate:

a Jim's taxable income

b how much income tax he should pay each year.

Use the diagram in Example 2.3 to help.

★ 8 Irene is a veterinary surgeon (vet).

- She earns a total annual income of £57 845 and her personal allowance is £10 000.
- She pays tax at the basic rate of 20% for the first £31 865 of taxable income and pays the higher rate of 40% on taxable income over £31 865.

Calculate:

a her taxable income

b how much income tax she should pay each year

c how much income tax she should pay each month.

9 David is a Member of Parliament (MP) and is registered blind.

- He earns a total annual income of £66 396.

- His personal allowance is £10 000 and he is also entitled to a blind person's allowance of £2230.

- He pays tax at the basic rate of 20% for the first £31 865 of taxable income and pays the higher rate of 40% on taxable income over £31 865.

Calculate:

a   his taxable income

b   how much income tax he should pay each year.

c   how much income tax he should pay each month.

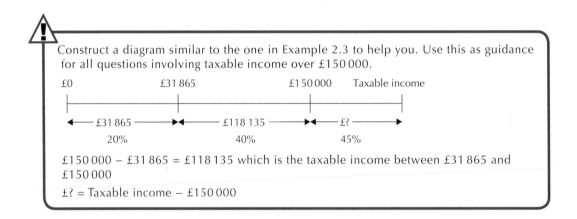

Construct a diagram similar to the one in Example 2.3 to help you. Use this as guidance for all questions involving taxable income over £150 000.

£150 000 − £31 865 = £118 135 which is the taxable income between £31 865 and £150 000

£? = Taxable income − £150 000

10 The chief executive of a major bank earns a total annual income of £182 550 and her personal allowance is £10 000.

She pays tax at the basic rate of 20% for the first £31 865 of taxable income, pays the higher rate of 40% on taxable income between £31 865 and £150 000, and an additional rate of 45% for taxable income earned over £150 000.

Calculate:

a   her taxable income

b   how much income tax she should pay each year.

11 A football manager earns a total annual income of £3.45 million. His personal allowance is £10 000.

He pays tax at the basic rate of 20% for the first £31 865 of taxable income, pays the higher rate of 40% on taxable income between £31 865 and £150 000, and an additional rate of 45% for taxable income earned over £150 000.

Calculate:

a   his taxable income

b   how much income tax he should pay each year

c   how much income tax he should pay each month.

12 Davina is an actor who worked on three different projects in one year. She records her income and tax rates as shown.

£36 255 for role in film

£123 827 for appearing in television series

£45 268 for performing in touring show

Personal allowance is £10 000

Tax paid at:

- 20% for the first £31 865 of taxable income
- 40% on taxable income between £31 865 and £150 000
- 45% for taxable income over £150 000

Calculate:

a her total annual income

b her taxable income

c how much income tax she should pay each year.

13 Joey works as a catering assistant earning the minimum wage of £6.31 per hour. He works 35 hours each week. His personal allowance is £10 000 and he pays tax at the basic rate of 20%.

a Calculate Joey's total annual income.

b Calculate his taxable income.

c Calculate how much income tax he should pay each year.

d Joey is involved in a car crash and loses his sight and is now entitled to a blind person's allowance of £2230. He returns to work and is still able to work full time. How much income tax should he now pay? Give a reason for your answer.

14 Freya is a senior social worker. She earns a total annual income of £31 718. Her personal allowance is £10 000 and she pays tax at the basic rate of 20%.

a Calculate:

i her taxable income

ii how much income tax she should pay each year.

b Express her income tax as a percentage of her total income.

National Insurance

National Insurance (NI) is also a tax on earnings. NI contributions build up your entitlement to certain state benefits, including unemployment, ill-health and the state pension.

Contributions are deducted from your wages by your employer.

The table shows NI rates for 2014–15. Note that the income figures are pre-tax.

As with income tax, NI rates will change each year. Look at the HMRC website to find out the current year's NI rates.

Rate	Weekly income	Annual income
0%	Less than £153	Less than £7956
12%	Between £153 and £805	Between £7956 and £41 860
2%	Above £805	Above £41 860

The 2014–15 rates are used for the Examples and Exercise questions in this chapter.

Example 2.4

Simon is a civil servant and earns a total annual income of £23 895.

a Using the table above calculate Simon's NI contribution for the year.

b Calculate his monthly NI contribution.

a £23 895 − £7956 = £15 939

> The rate of 12% is paid for taxable income between £7956 and £41 860. Find the amount which is paid at 12%. The diagram represents this information.

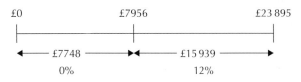

Annual NI contribution is 12% of £15 939 = £1912.68

b Monthly NI contribution is £1912.68 ÷ 12 = £159.39

Example 2.5

Marie is an accountant and earns £1057.70 per week. Use the table above to calculate her weekly NI contribution.

£1057.70 − £805 = £252.70

> Use a diagram to help you work out the rates to apply to different amounts.

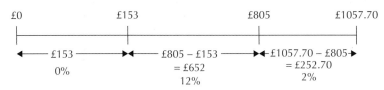

12% of £652 = £78.24

2% of £252.70 = £5.05 (2 d.p.)

Weekly NI contribution is £78.24 + £5.05 = £83.29

Exercise 2C

1 Calculate the annual NI contribution for each person.

    a Jess, a trainee lawyer, earns a total annual income of £16 650.

    b Norman, an electrician, earns a total annual income of £23 684.

    c Sylvia, a website designer, earns a total annual income of £21 437.

   d   Sergei, a television executive, earns a total annual income of £102 354.

   e   Gina, a doctor, earns a total annual income of £81 423.

   f   Frances, a part-time school crossing guard (also called a 'Lollipop lady'), earns a total annual income of £6987.

2   Calculate the weekly NI contribution for each person.

   a   Susan, an architect, earns a weekly income of £1307.65.

   b   Graham, a shoe shop sales assistant, earns a weekly income of £403.85.

   c   Jake, a music teacher, earns a weekly income of £657.85.

   d   Elspeth, a zookeeper, earns a weekly income of £308.27.

3   Joy works as a cleaner. She earns £6.52 per hour and works a 35-hour week.
    Calculate her weekly NI contribution.

4   Frank works as a gardener. He earns a total annual income of £17 297.
    Calculate:

   a   his annual NI contribution.

   b   his monthly NI contribution.

5   Alana is a soldier in the regular army. Her weekly NI contribution is £16.32. Calculate her weekly wage.

> £16.32 is 12% of the original amount. Find 100%, then add £153 (0%).

6   Alan is a mechanic. His annual NI contribution is £1759.68. Calculate his weekly wage.

★ 7   Sara is a pharmacist and earns a total annual income of £37 215.

   a   Calculate her annual NI contribution.

       Sara receives a pay rise of 4%.

   b   i   Calculate her new annual NI contribution.

       ii   By how much does her annual NI contribution rise?

Pension contribution

Pension contribution is the amount deducted from your salary which goes towards your pension scheme.

Employers set up a pension scheme so that you receive regular payments when you retire. Monthly payments towards this scheme are usually calculated as a percentage of the employee's gross pay. Each scheme is different but pension contributions are generally around 5% of the gross salary.

Payslip

Anyone who is in employment will receive a payslip from their employer showing all their earnings and deductions for the latest week or month.

Example 2.6

This is Nathan's payslip for a particular month. He works as an office manager.

Personal details

McGregor and Jones LLP						
Name N Hamilton		**Employee number** 7345		**NI number** HG018204B	**Month number** 5	
Payments	**Value**	**Deductions**	**Value**			
Basic pay Overtime Bonus	£1955.58 £32.24 £40	Income tax National Insurance Pension	£233.78 £157.19 £97.78			
Gross pay	£	**Total deductions**	£	**Net pay**	£	

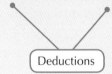 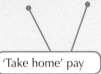

Basic pay plus other income   Deductions   'Take home' pay

a Calculate Nathan's gross pay.

b Calculate his total deductions.

c Calculate his net pay.

d Express his net pay as a percentage of his gross pay (correct to 1 d.p.).

a Gross pay is £1955.58 + £32.24 + £40 = £2027.82

b Total deductions is £233.78 + £157.19+ £97.78 = £488.75

c Net pay is £2027.82 − £488.75 = £1539.07

d Net pay as percentage of gross pay is $\dfrac{1539.07}{2027.82} \times 100 = 75.9\%$

Exercise 2D

1 Jane is a care assistant. Here is her payslip this week.

Care 4 U						
Name J Brown		**Employee number** 4368		**NI number** JE129574K	**Week number** 23	
Payments	**Value**	**Deductions**	**Value**			
Basic pay Overtime	£319.65 £45.67	Income tax National Insurance Pension	£15.93 £20.48 £15.98			
Gross pay	£	**Total deductions**	£	**Net pay**	£	

Calculate Jane's:

a gross pay

b total deductions

c net pay.

2 Tom is a personal trainer. Here is his payslip this month.

Gyms R Us						
Name T Hart		**Employee number** 536		**NI number** NH937920P	**Month number** 7	
Payments	**Value**	**Deductions**	**Value**			
Basic pay Bonus	£2385.58 £25	Income tax National Insurance Pension	£319.78 £208.79 £143.13			
Gross pay	£	**Total deductions**	£	**Net pay**	£	

a Calculate Tom's gross pay.

b Calculate his total deductions.

c Calculate his net pay.

d Express his pension contribution as a percentage of his gross pay (correct to the nearest whole number).

3 Hattie is the Principal Teacher of PE in a secondary school.

Beaton High School					
Name H Mansfield		**Employee number** 7301		**NI number** MD969621C	**Month number** 1
Payments	**Value**	**Deductions**	**Value**		
Basic pay	£4010	Income tax National Insurance Pension Professional fee	£604.67 £200.52 £45		
Gross pay	£4010	**Total deductions**	£1198.28	**Net pay**	£

a Calculate her NI contribution.

b Calculate her net pay.

c Express her net pay as a percentage of her gross pay (correct to 1 d.p.).

★ 4 Ruth is a chef. She earns a total annual income of £24 112.

a Calculate her monthly basic pay.

b Her personal allowance is £10 000 and she pays tax at the basic rate of 20%.
Calculate:
  i her taxable income
  ii how much income tax she should pay each year
  iii how much income tax she should pay each month.

c Using the table on page 31 calculate:
  i her annual NI contribution
  ii her monthly NI contribution.

d Ruth's employers offer a pension scheme. Her contribution is 5% of her monthly basic pay.
Calculate her monthly pension payment.

e   Using your answers to parts **a–d**, copy and complete Ruth's payslip below.

Walker Lodge Hotel					
Name R Potter		**Employee number** 539		**NI number** YQ849371H	**Month number** 5
Payments	**Value**	**Deductions**	**Value**		
Basic pay	£	Income tax National Insurance Pension	£ £ £		
Gross pay	£	**Total deductions**	£	**Net pay**	£

5   Ella works at a supermarket and gets paid £6.85 per hour. She works a 30-hour week.

a   Calculate her weekly wage (basic pay).

b   Her personal allowance is £10 000 and she pays tax at the basic rate of 20%.

⚠ Calculate the annual NI Contribution then divide by 12.

Calculate:

i   her annual salary

ii   her taxable income

iii   how much income tax she should pay each year

iv   how much income tax she should pay each week.

c   Using the table on page 32 calculate:

i   her weekly NI contribution.

d   Ella's employers offer a pension scheme. Her contribution is 4% of her weekly basic pay.

Calculate her weekly pension contribution.

e   Using your answers to parts **a–d**, copy and complete Ella's payslip below.

Save 'n' Spend					
Name E Wilson		**Employee number** 52384		**NI number** NH103872F	**Week number** 31
Payments	**Value**	**Deductions**	**Value**		
Basic pay	£	Income tax National Insurance Pension	£ £ £		
Gross pay	£	**Total deductions**	£	**Net pay**	£

6   George works in a contact centre (formerly known as a call centre) and gets paid £10.50 per hour. He is registered blind.

He works 8 a.m. to 6 p.m. on Mondays, Tuesdays and Wednesdays and gets an hour off (unpaid) for lunch each day. He also works 10.30 a.m. to 2.30 p.m. on Thursdays and Fridays.

a   i   How many hours does he work in a week?

ii   Calculate his weekly wage (basic pay).

George's personal allowance is £10 000 and he is also entitled to a blind person's allowance of £2230. He pays tax at the basic rate of 20%.

b   Calculate:

i   his annual salary

ii   his taxable income

  iii how much income tax he should pay each year

  iv how much income tax he should pay each week.

c Using the table on page 32 calculate:

  i his weekly NI contribution.

d George's employers offer a pension scheme. His contribution is 5.5% of his weekly basic pay. Calculate his weekly pension contribution.

e Using your answers to parts **a–d**, copy and complete George's payslip below.

Telecommunications International						
Name G MacRae		**Employee number** 7269			**NI number** AB926792T	**Week number** 12
Payments	**Value**	**Deductions**		**Value**		
Basic pay	£	Income tax National Insurance Pension		£ £ £		
Gross pay	£	**Total deductions**		£	**Net pay**	£

7 Eric works as a footballer. He earns a total annual income of £3.4 million.

a Calculate his monthly wage (basic pay).

b Eric's personal allowance is £10 000. He pays tax at the basic rate of 20% for the first £31 865 he earns, pays the higher rate of 40% on taxable income between £31 865 and £150 000 and an additional rate of 45% for taxable income earned over £150 000.

 Calculate:

  i his taxable income

  ii how much income tax he should pay each year

  iii how much income tax he should pay each month.

c Using the table on page 32 calculate:

  i his annual NI contribution.

  ii his monthly NI contribution.

d Eric's employers offer a pension scheme. His contribution is 1.5% of his monthly basic pay. Calculate his monthly pension contribution.

e Using your answers to parts **a–d**, copy and complete Eric's payslip below.

Ayton United Ltd						
Name E Milton		**Employee number** 728			**NI number** JF192749A	**Month number** 3
Payments	**Value**	**Deductions**		**Value**		
Basic pay	£	Income tax National Insurance Pension		£ £ £		
Gross pay	£	**Total deductions**		£	**Net pay**	£

f Express his net pay as a percentage of his gross pay.

(GO!) Activity

This budgeting activity combines the work you have done in Chapters 1 and 2.

1 Choose a job, perhaps the job you would like to do when you are older.

> ⚠ Try searching on the PlanIT Plus

Research the **starting salary** for this job.

a Divide the salary by 12 to calculate your monthly income.

> ⚠ The starting salary is the salary you will earn when you are first qualified before you have gained experience or promotion.

The table shows the different rates of tax that are applied to different amounts of taxable income.

Rate (%)	Taxable income
0	up to £10 000
20	up to £31 865
40	£31 865–£150 000
45	above £150 000

b Use the information in the table to calculate:

  i the taxable income, assuming that you are only entitled to a personal allowance of £10 000

  ii how much income tax you should pay each year

  iii how much income tax you should pay each month.

2 a Use the information about NI contributions in the table on page 32 to calculate:

  i your annual NI contribution

  ii your monthly NI contribution.

  b Assume that your employer offers a pension scheme. Calculate your monthly pension contribution as 5% of your monthly basic pay.

3 Calculate your:

  a gross pay

  b total deductions

  c net pay.

4 a Draw up an expenditure table like the one below or create one using a spreadsheet program.

Expenditure		
Fixed expenses	Rent or mortgage	£
	Council tax	£
	Gas and electricity	£
	Telephone	£
	Travel	£
	TV licence	£

Variable expenses	Food/supermarket	£
	Clothes	£
	Going out	£
	Other (e.g. gym, music, books, entertainment)	£
	Savings	£
Total expenditure		£

If you have drawn up a paper version of this expenditure table, it is recommended that you use a pencil when writing in the figures as you may have to alter your expenditure as you work through the activity.

b Now decide where you will live. Assume that you have to move out of your parents' home. Here are the likely costs.

Studio flat	**1 bedroom flat**	**2 bedroom flat**
£250 rent	£400 rent	£600 rent
£280 mortgage	£425 mortgage	£625 mortgage
3 bedroom house (in the suburbs)	**3 bedroom cottage** (in the country)	**6 bedroom house** £2000 rent
£800 rent	£680 rent	£1800 mortgage
£750 mortgage	£650 mortgage	

Once you have decided where you will live, add this cost to your expenditure table.

There are different consequences of renting and buying. A mortgage will give you 'something for your money' in the long term but is more expensive in the short term as you need to build up an initial deposit for this. You are also responsible for all upkeep and repairs.

c Your choice of accommodation affects your council tax and gas/electricity expenses. Add these to your expenditure table.

Studio flat	**1 bedroom flat**	**2 bedroom flat**
Gas/electricity £40	Gas/electricity £60	Gas/electricity £80
Council tax £65	Council tax £80	Council tax £100
3 bedroom house	**3 bedroom cottage**	**6 bedroom house**
Gas/electricity £120	Gas/electricity £130	Gas/electricity £220
Council tax £140	Council tax £140	Council tax £250

d You also need to consider how you will travel to work. Travel costs often depend on your choice of accommodation as flats are often closer to workplaces than houses. Public transport options may be limited in some locations. If you have a car, then there are expenses such as car loan, tax, insurance and repairs, as well as weekly fuel costs.

These costs are a guideline only. Each house or flat will have a different cost for gas/electricity and council tax costs depend on the actual location.

(continued)

e   Use the following table of monthly costs to choose how you will travel.

	New car (£)	Used car (£)	Old car (£)	Train (£)	Bus (£)
Studio flat	200	150	100	80	60
1 bedroom flat	200	150	100	80	60
2 bedroom flat	200	150	100	80	60
3 bedroom house	250	200	130	130	70
3 bedroom cottage	300	250	200	Not an option (no station nearby)	150 (but infrequent service)
6 bedroom house	250	200	130	130	70

5   Food and certain items such as soap and toilet paper are considered essential items. Here is an example of a weekly shopping list.

Chicken drumsticks 450 g   Packet of muesli (500 g)

1 kg potatoes   6 eggs

Packet of mixed salad 150 g   Bread (small)

Packet of 2 boneless salmon fillets   Tea bags (40)

Baked beans (420 g)   1 litre of semi-skimmed milk

1 tin chick peas   1 litre of fruit juice

Packet of rice   4-pack of soap

Packet of pasta   Washing powder (1.2 kg)

1 kg bananas   1 bottle washing up liquid

Tub of margarine (500 g)   Toilet rolls (4)

Use this list and research the cost of these items at a supermarket of your choice.

You should find that the amount comes to approximately £30.

Multiply your answer by 4 for an approximate food/supermarket expense and add this to your expenditure table.

6   Now you need to decide how you will spend the rest of your income. Choose from the following.

Telephone	£10	Internet	£15
Clothes	£20 per month	Clothes	£40 per month
Clubbing	£15	TV licence	£16
Mobile phone	£25	Meal out	£25
Clothes	£60 per month	Cinema	£7
Gym membership	£35	DVD	£8
Music (downloads or CDs)	£12	Games console video game	£27
Presents and cards	£15		

Add your choices to your expenditure table.

7   a   Use your expenditure table and calculate your savings.

    b   Now reassess your table. Do you need to make any changes in order to generate some savings? Do you have to alter your choice of accommodation? For example, could you and a friend share a 2 bedroom flat? What about your variable expenses – are there ways of cutting back?

    c   Compare your expenditure table with others in the class. How are their decisions similar to yours and how do they differ?

---

* I can investigate and interpret income for different circumstances and career choices including basic pay, gross/net pay, incentive payments, benefits and allowances.
  ★ Exercise 2A Q1

* I can investigate and interpret deductions for different circumstances and career choices including income tax.
  ★ Exercise 2B Q8

* I can investigate and interpret deductions for different circumstances and career choices including National Insurance contributions.
  ★ Exercise 2C Q7

* I can investigate and interpret deductions for different circumstances and career choices including pension contributions.
  ★ Exercise 2D Q4

For further assessment opportunities, see the Case Studies for Unit 1 on pages 136–140.

3 Determining the best deal, given three pieces of information

This chapter will show you how to:

- compare at least three products, given three pieces of information on each.

You should already know:

- how to compare at least three products, given two pieces of information on each
- how to interpret data from a table with at least four categories of information
- how to calculate a percentage of a quantity.

Determining the best deal

The cheapest deal may not always be the best deal. In this chapter you will learn how to extract information from a table and make decisions on the best deal in terms of cost as well as quality of service or product or other relevant factors.

This is a skill you will need to use in life because it is important to 'shop around' for the best deal rather than accepting the first deal offered.

Example 3.1

Jonny wants to buy new windows and doors for his house.

He needs to replace two windows in his living room, the front door and a set of patio doors.

The company PVC Queen sells windows and doors.

Its prices for the dimensions Jonny needs are given below.

Window	Cost	Front door	Cost	Patio door	Cost
Style 1	£623	Flat panel door	£395	Inline doors	£877
Style 2	£667	Clear glazed door	£599	French doors	£925
Style 3	£703	Clear crystal door with Georgian panel	£649	French rock doors	£1425

In addition, the installation of each window will cost 20% of the price, and the installation of each type of door will cost 25% of the price.

a What is the cheapest cost of two windows, one door and one set of patio doors, including installation cost?

b Jonny has a budget of £3600.

By investigating the combinations of windows, doors and patio doors, find the maximum he can spend within the £3600 limit.

a Windows (Style 1)

The cheapest window is Style 1.

Cost is £623 × 2 = £1246

Installation is 20% of £1246 = 0.2 × £1246 = 249.20

Total cost is £1246 + £249.20 = £1495.20

Front door (Flat panel door)

Cost is £395

Installation is 25% of £395 = 0.25 × £395= £98.75

Total cost is £395 + £98.75 = £493.75

Patio doors (Inline doors)

Cost is £877

Installation is 25% of £877 = 0.25 × £877 = £219.25

Total cost is £395 + £98.75 = £1096.25

Cheapest cost is £1495.20 + 493.75 + 1096.25 = £3085.20

b Investigation leading to:

Windows (Style 3)

Cost is £703 × 2 = £1406

Installation is 20% of £1406 = 0.2 × £1406 = £281.20

Total cost is £1406 + 281.20 = £1687.20

Front door (Clear crystal door with Georgian panel)

Cost is £649

Installation is 25% of £649 = 0.25 × £649 = £162.25

Total cost is £649 + £162.25 = £811.25

Patio doors (Inline doors)

Cost is £877

Installation is 25% of £877 = 0.25 × £877 = £219.25

Total cost is £877 + £219.25 = £1096.25

Cost closest to budget of £3600 is £1687.20 + 811.25 + 1096.25 = £3594.70

Exercise 3A

★ 1 Sarah is planning a weekend trip with two nights' accommodation.

She would like to visit Bristol, York or Brighton.

She gets the following information about three-star accommodation from the internet.

Location	Accommodation	Price per night (£)
Bristol	Avon Hotel	87
	Clifton Trees Hotel	53
	Hotel Brunel	45
York	Castle Hotel	106
	Ouse Hotel	69
	Hotel Nordic	95
Brighton	Pavilion Hotel	94
	Lanes Hotel	56
	Hotel Seattle	59

She would also like to visit tourist attractions while on the trip. Here are the attractions she would like to visit.

Location	Attraction	Ticket price (£)
Bristol	Zoo Gardens	15.50
	Science Centre	12.90
	Brunel SS Great Britain	12.95
York	Castle Howard	14.00
	Jorvik Viking Centre	9.25
	York Minster	10.00
Brighton	Royal Pavilion	10.50
	Brighton Wheel	8.00
	Seal Life Brighton	14.10

Sarah lives in Edinburgh. The cost of travel to each place is listed below.

Location	Cost of a single (one-way) ticket	Ticket price (£)
Bristol	Flight to Bristol Airport	42.00
	Coach from Bristol Airport to city centre	7.00
York	Train to York city centre	36.00
Brighton	Flight to Gatwick Airport	37.00
	Train from Gatwick Airport to Brighton	9.50

a Sarah decides to choose the cheapest accommodation, visit the two cheapest tourist attractions and will travel either by train or air.

Which place is the cheapest to visit? Give a reason for your answer.

b She has a budget of £250.

She would like to stay in the Pavilion Hotel in Brighton and visit all three tourist attractions.

   i   Is Sarah's budget enough to pay for this? Use your calculations to justify your answer.

   ii Name one change she could make to the above decisions in order to ensure she is under budget.

2 A recipe for Chicken Chasseur lists these ingredients:

4 chicken breasts

50 ml olive oil

2 onions

250 g mushrooms

25 ml tomato purée

300 ml white wine

400 g can chicken consommé

4 tomatoes

Jo researches the cost of the ingredients in three different supermarkets.

Ingredient	Albertsons	Everyday	Festival Foods
4 chicken breasts	2 for £3.80	4 for £6.89	2 for £3.90
50 ml olive oil	500 ml bottle £2.00	250 ml bottle £1.25	250 ml bottle £1.45
2 onions	16p per onion	25p per onion	29p per onion
250 g mushrooms	£2.15 per kg	79p for 250 g pack	300 g pack for £1.15
25 g tomato purée	200 g tube 48p	200 g tube 46p	200 g tube 48p
300 ml white wine	750 ml bottle £3.84	750 ml bottle £3.75	750 ml bottle £4.00
400 g can chicken consommé	400 g can £1.69	400 g can £1.73	400 g can £1.68
4 tomatoes	17p per tomato	14p per tomato	18p per tomato

a If she buys all the ingredients in the same shop, which shop will be the cheapest to buy all the ingredients Jo needs for the recipe?

b If she buys the cheapest deal for each ingredient:
   i   how much will she spend on each ingredient?
   ii how much will she spend in each shop?
   iii how much will this cost her altogether?
   iv express your answer to part **b iii** as a percentage of your answer to part **a**.

3 Harry and Meg want to go on holiday to Alicante in Spain.

Rather than buy a package holiday they decide to book flights and accommodation independently.

For the dates they want to travel they find three companies that offer flights from Glasgow to Alicante. The table on the next page shows the cost per person on each flight.

They want to check in 2 bags (1 each) each way.

Airline	Flight out		Return flight		Cost of baggage
	Time	Cost (£)	Time	Cost (£)	
Jet off	0700	£104	1300	£119	Included in cost of flight
Flyaway	1200	£126	1845	£90	£20 per bag each way
Belle Air	1830	£81	2230	£105	£15 per bag each way

They also need to pay for accommodation. They find three hotels which offer room only, bed and breakfast, and full board (all meals provided). The prices below show the cost per person for 7 nights' accommodation.

Hotel	Accommodation type	Price per person (€)
Hotel Melia	Room only	389
	Bed and breakfast	456
	Full board	524
Hotel Maris	Room only	335
	Bed and breakfast	453
	Full board	570
Hotel Explanada	Room only	315
	Bed and breakfast	465
	Full board	598

The exchange rate is £1 = €1.21.

They also need to buy travel insurance. The table below shows the cost of travel insurance to different destinations for different periods of time

Length of holiday	Cost per person for travel insurance (£)			
	UK	Europe	USA/Canada	Rest of the world
1–4 days	5.50	6.75	8.35	9.15
5–12 days	6.85	8.25	9.45	10.25
13–24 days	7.35	9.65	10.95	11.55
25+ days: add £0.30 per day to the cost for 13–24 days				

a   Calculate the cost (in £) of the cheapest holiday they can purchase. Include flights, accommodation and travel insurance.

b   Calculate the cost (in £) of the most expensive holiday they can purchase.

> ⚠ Make sure that you use the exchange rate to calculate the cost in pounds, and remember to round your answer to two decimal places.

c   If they decide to travel with Flyaway (because their times are the most convenient) and choose accommodation offering bed and breakfast, what is the cost of the cheapest holiday they can buy? Which hotel will they stay in?

d   They have a budget of £1300. Investigate the combinations of flights and accommodation, including the cost of travel insurance, to find the maximum they can spend within their £1300 limit. Which options can they choose?

4   The table below shows the monthly costs of different companies' TV, broadband and phone packages.

Company	Cost of TV (£)	Number of TV channels	Cost of broad-band (£)	Cost of line rental (£)	Upfront costs (£)	Special offers
Frontier Communications	14.50	316	7.50	15.40	None	First 6 months TV reduced to £7.50
Comcast	14.00	80	7.25	15.45	41.95	First 6 months broadband reduced to £5
Matrix Technology	10.95	82	5.65	15.75	25.00	No offer
Closecall UK	9.95	137	6.90	15.60	42.00	£25 supermarket giftcard
Wireless Communications	18.00	382	9.00	15.50	None	First 6 months broadband and TV reduced by 50%

a   Calculate the cost for the first year of the package offering the most TV channels.

b   By calculating the cost of the first year of each package, which company offers the best deal?

c   Special offers and upfront costs only apply in the first year of each package. Calculate the cost of the second year of each package. Which company offers the best deal in the second year?

d   Which company offers the best deal for the first 2 years combined?

5   Ticket prices for different sports at a major sporting event are shown in the table.

Annie and Angus live in Fife. They decide to take their three children Louise (aged 15), Sam (aged 7) and John (aged 4) as well as Annie's mother Morag (aged 68).

As well as paying for the tickets, they need to pay for travel to and from the event which takes place in Glasgow.

Sport	Adult ticket (£)*	Child ticket (£)**
Athletics	21	15
Badminton	14	8
Cycling	16	10
Gymnastics	18	11
Netball	15	8
Swimming	20	13

*Concession tickets (aged 60 and over): 75% of adult ticket
**Child tickets: for children aged 5–14 years
**Under 5s: free

Mode of transport	Adult fare (£)	Concession fare (£)	Child fare (£)	
			5–15 years	Under 5 years
Train	£20.30 return	$\frac{2}{3}$ of adult fare	$\frac{1}{2}$ of adult fare	free
Coach	£7.80 each way	free	£7.80 each way	Under 2: free 3–4 years: $\frac{1}{2}$ of adult fare

Each stadium offers the same prices for the fast food kiosk.

Annie and Angus usually order chicken burgers, Morag usually orders a veggie burger, Louise usually orders a beef burger and the two smaller children (Sam and John) order the children's meals.

Meal deals	Cost (£)
Beef burger, fries and drink	4.95
Chicken burger, fries and drink	4.35
Veggie burger, fries and drink	4.05
Children's burger meal	1.95

a   If they choose to go to the three cheapest sports (a different sport each day), calculate the minimum cost of three days at the event, assuming that they also need to pay for travel and a meal on each day.

b   If they choose to go to all of the events and take the train each day, calculate the minimum cost of their three days at the event, assuming that they also need to pay for travel and a meal on each day.

6   The table below shows the cost of different companies' mobile phone packages for the same handset.

Company	Cost of handset (£)	Monthly handset payment (£)	Monthly usage cost (£)	Minutes/texts/data allowance	Cost of extra texts	Cost of extra minutes
A	284.99	15	12.99	1000 min Unlimited texts 500 MB data	n/a	9p per min
B	274.99	14	13.50	250 min Unlimited texts Unlimited data	n/a	10p per min
C	245.99	17	11.00	250 min 500 texts Unlimited data	12p per text	8.5p per min
D	free	16	21.00	Unlimited min 1000 texts 1 GB data	14p per text	n/a
E	free	16	36.00	Unlimited min Unlimited texts 2 GB data	n/a	n/a

Gillian wants to buy this type of phone.

On average she uses her phone for $7\frac{1}{2}$ hours of calls and sends 1200 texts each month.

a i Calculate the cost of each mobile phone package for the first year, assuming her usage remains the same each month.

  ii Which package gives the best deal? Give a reason for your answer.

b i If she reduced her usage to $4\frac{1}{2}$ hours of calls and 800 texts per month, which company would give the best deal for the first year?

  ii If she wants a minimum of 2 GB of data allowance as well as $4\frac{1}{2}$ hours of calls and 800 texts per month, which company gives the best deal for the first year?

7 Marco wants to take his three grandchildren to the theatre in Edinburgh to see a popular show. Marco's grandchildren are aged 16, 13 and 9. The table below shows the price of tickets for different types of seats at different performances.

Performance	Seating area				
	Front stalls (£)	Rear stalls (£)	Circle (£)	Front balcony (£)	Rear balcony (£)
Tuesday, Wednesday or Thursday evening	50	45	55	40	30
Wednesday or Saturday matinée	53	47	60	43	32
Friday or Saturday evening	55	50	60	45	35

*Children under 16: £10 discount
**1.5% booking fee added to the cost of each ticket

They will eat at the nearby Italian restaurant which offers different deals according to the time of week.

Menu	Cost for 2 courses (£)	Cost for 3 courses (£)
Lunch menu Monday to Friday	8.95	11.95
Lunch menu Saturdays and Sundays	10.50	12.95
Pre-theatre menu Monday to Friday	10.95	13.95
Pre-theatre Saturday and Sundays	11.50	15.50

Add £2 per person to the cost of a meal for drinks.
Add 10% to the total cost of the meal for a tip to the waiter.

They will also have a snack at the theatre during the interval, and the table lists snack prices.

a Calculate the cheapest cost of attending the theatre on each of the following days, assuming Marco buys the cheapest interval snack for himself

Interval snack	Price (£)
Ice cream	2.50
Ice lolly	2.15
Packet of chocolate sweets	2.99
Packet of fruit sweets	2.69

and each grandchild and takes them to the restaurant for a 2-course meal before the show:

  i Wednesday matinée

  ii Thursday evening

  iii Saturday matinée

  iv Saturday evening.

b Marco has a budget of £250.

  Investigate combinations of ticket prices, menus and snacks (one per grandchild) to find the maximum Marco can spend within his £250 limit.

8 A make of car is sold with different specifications and optional extras. The table below shows the basic price of each car and the cost of optional extras.

| Type | Number of doors | Basic price (£) | LED daytime running lights (£) | In-car entertainment (£) | Paint | |
					White (£)	Metallic (£)
Basic	3	6150	180	175	150	400
Regular	3	7500	180	175	150	400
Deluxe	3	8100	190	185	170	420
Regular	5	7800	180	175	150	400
Deluxe	5	8440	190	185	170	420
Special	5	8680	200	190	190	440

Value added tax (VAT) at 20% is to be added to each of the prices.

A buyer must choose either white paint or metallic paint.

Bob has decided to buy this make of car. He has a budget of £10 800.

a Explain why he cannot afford to buy the Deluxe 5-door car with optional extras (LED lighting and in-car entertainment) and metallic paint.

b Investigate the combination of type of car and optional extras to find the maximum he can spend within his £10 800 limit.

9 An optician's shop sells glasses in a variety of frames, lenses and cases. The tables show the prices of different options.

a How much would it cost to buy a pair of glasses with:

  i designer frame, varifocal lenses and a black zipper case

  ii semi-rimless frame, elite lenses and a hard case?

Frame	Price (£)
Basic	£45
Stainless steel/flexible hinges	£85
Semi-rimless	£99
Designer	£125
Rimless	£169

b The shop has a special offer: buy one pair, get a second pair half price. This means that the cheaper pair (including case) is sold at half price.

What is the cost of buying a pair of glasses with rimless frame, bifocal lenses and a chrome case and a pair of glasses with a stainless steel/flexible hinges frame, standard lenses and a black zipper case?

c You have a budget of £170 and want to buy one pair of glasses. Investigate the combination of frames, lenses and cases to find the maximum you can spend within your £170 limit.

Lens	Price (£)
Standard	£49
Varifocal	£79
Elite	£109
Bifocal	£40

Case	Price (£)
Black zipper	£5.99
Hard case	£7.50
Chrome case	£8.99

Make the Link

There are many comparison websites to help you find the best deal possible, for example, 'Go Compare', 'Compare the Market', 'Money Supermarket', 'Confused' (car insurance) and 'Uswitch' (gas and electricity). These websites claim to be impartial and are a good way of working out good deals within your budget.

Activity

You decide to go on holiday with a friend. Choose a destination, for example, Majorca, Kos or Gran Canaria, and research (using the internet or brochures from a travel agent) different combinations of costs for a set time of 7 days for the first week in July.

Set yourself a budget (e.g. £600 each) and investigate the costs of different types of accommodation, flights, meal options (room only, half-board or all-inclusive) and holiday activities.

Compare your answers with others and find the best deal within your given budget.

- I can compare at least three products, given three pieces of information on each. ★ Exercise 3A Q1

For further assessment opportunities, see the Case Studies for Unit 1 on pages 136–140.

4 Converting between several currencies

This chapter will show you how to:

- convert between currencies involving the use of at least three currencies in a multi-stage task.

You should already know:

- how to convert between pounds sterling and other currencies
- how to compare costs between two different currencies.

Converting between currencies

When you go abroad you need to have money in the currency of the country or countries you are visiting. You do this by exchanging British currency (pounds sterling) for the currency you want.

An **exchange rate** is the rate at which one unit of the currency of one country can be exchanged for another country's currency. Exchange rates change daily because they are influenced by a wide variety of factors such as interest rates, inflation and the economies of individual countries.

When currency is being exchanged, the bank, travel agent or Post Office (whoever is actually exchanging your money for the other currency) might charge a **commission** for performing the service. Sometimes this is a flat fee, and sometimes it is a percentage of the amount being exchanged.

Exchange rate calculations

To exchange pounds sterling (£) into a foreign currency:

amount in new currency = amount to be exchanged (in £) × exchange rate

To exchange foreign currency into pounds sterling (£):

amount in £ = amount to be exchanged (in foreign currency) ÷ exchange rate

The questions in this chapter involve more than one foreign currency and use the above methods to solve questions.

Use the exchange rates in the table opposite to answer questions in the examples and exercise in this chapter, unless you are advised otherwise.

Different currencies use different symbols. The symbol for the dollar is $, so the currencies of Australia and Canada, for example, which use the dollar, are distinguished by the letters AUD and CAD before the symbol.

Country	Currency (symbol)	Exchange rate
Australia	Australian dollar (AUD$)	1.79
Canada	Canadian dollar (CAD$)	1.74
China	Chinese yuan renminbi (CNY¥)	10.35
European countries in Eurozone	euro (€)	1.21
India	Indian rupee (₹)	102.27
Japan	Japanese yen (JPY¥)	167.71
Switzerland	Swiss franc (CHF)	1.48
South Africa	South African rand (R)	16.67
Turkey	Turkish lira (₺)	3.31
USA	American dollar (USD$)	1.64

⚠ The Eurozone is the collection of European countries that use the euro for their currency. Not all European countries are in the Eurozone. The countries which use the euro as their currency are: Belgium, Germany, Ireland, Spain, France, Italy, Luxembourg, Netherlands, Austria, Portugal, Finland, Greece, Slovenia, Cyprus, Malta, Slovakia and Estonia.

Example 4.1

Mark and Karen travel to the USA and Canada. For their spending money they exchange £600 into US dollars and £450 into Canadian dollars.

a   How much of each currency do they get?

b   They spend $742 in the USA and $681 in Canada. On return to the UK they exchange their remaining money back to pounds sterling.

How much (in total) do they receive once they have exchanged their remaining money?

a   US dollars: £600 × 1.64 = USD$984

Canadian dollars: £450 × 1.74 = CAD$783

b   US dollars: USD$984 − USD$742 = USD$242

GB pounds: USD$242 ÷ 1.64 = £147.56   ●————————————(Round to 2 d.p.)

Canadian dollars: CAD$783 − CAD$681 = CAD$102

GB pounds: CAD$102 ÷ 1.74 = £58.62

Total = £147.56 + £58.62 = £206.18

Example 4.2

Cindy wants to exchange £250 into euros so she has spending money for a weekend break in Amsterdam. She does not want to spend any more than £250.

Two travel agents offer different deals.

Travel agent	Exchange rate	Commission
1	£1 = €1.19	No commission charged
2	£1 = €1.23	3% commission charged

Which travel agent offers the better deal? Give a reason for your answer.

(continued)

Travel agent 1: 250 × 1.19 = €297.50

Travel agent 2: 250 × 1.23 = €307.50

Commission: 3% of 307.50 = 0.03 × 307.50 = €9.23 (2 d.p.)

Total due: €307.50 − €9.23 = €298.27

Cindy should go to Travel agent 2 because she will get €298.27 which is €0.77 more than Travel agent 1 offers.

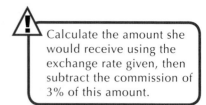

Calculate the amount she would receive using the exchange rate given, then subtract the commission of 3% of this amount.

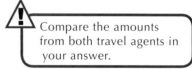

Compare the amounts from both travel agents in your answer.

Exercise 4A

Use the currencies in the table on p55 to answer all questions unless advised otherwise.

1 Simon and Debbie travel to Japan, India and Australia. For their spending money they exchange £800 into Japanese yen, £600 into Indian rupees and £1000 into Australian dollars.

 a How much of each currency do they receive?

 b They spend JPY¥112 400 in Japan, ₹55 200 in India and AUD$1645 in Australia.

 On return to the UK they exchange their remaining money back to pounds sterling.

 How much (in total) do they receive once they have exchanged their remaining money?

★ 2 Sam and Anna are planning a trip to Munich (Germany), Bern (Switzerland) and Milan (Italy). They research accommodation and find the following prices.

Location	Name of hotel	Cost per person per night
Munich	Hotel Europa München	€78
Bern	Hotel Alpen	CHF155
Milan	Hotel Michelangelo	€95

They decide to add a tip of £10 per night (between them) onto all accommodation costs.

 a They book 2 nights in Munich. How much (including tip) will this cost them in euros?

 b They book 3 nights in Bern. How much (including tip) will this cost them in Swiss francs?

 c They book 2 nights in Milan. How much (including tip) will this cost them in euros?

 d Convert your answers to parts a–c into pounds sterling and calculate the total cost of accommodation for their holiday.

3 A chain of fast food restaurants has restaurants in London (UK), Sydney (Australia), Tokyo (Japan), Paris (France) and Cape Town (South Africa).

Meal	Cost in each city				
	London (£)	Sydney (AUD$)	Tokyo (JPY¥)	Paris (€)	Cape Town (R)
Quarter pound burger	2.59	3.99	440	3.60	30.00
Chicken burger	2.45	4.19	400	3.40	25.00
9 chicken nuggets	2.99	5.00	425	4.30	35.00
Cheese burger	0.99	1.29	140	1.45	15.00
Regular fries	1.09	1.49	240	1.55	16.30
Regular drink	1.05	1.25	125	1.49	11.50

a   If you wanted to buy 2 quarter pound burgers, a chicken burger, 9 chicken nuggets, 3 portions of fries and 4 drinks, how much would you pay in:

   i   London   ii   Sydney   iii   Tokyo

   iv   Paris   v   Cape Town?

b   i   Convert your answers in part **a** to pounds sterling.

   ii   Which country is the cheapest for your order?

   iii   What is the difference in cost for the order between the cheapest and the most expensive country?

c   Why is there a difference in cost between the menus in the different countries?

★   **4**   A camera is sold in five different countries.

Country	Camera price
Germany	€195
USA	USD$280
India	₹16 100
Japan	JPY¥27 500
Switzerland	CHF260

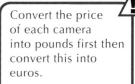

Convert the price of each camera into pounds first then convert this into euros.

a   Convert the cost of the camera in each country into euros.

b   In which country is the camera cheapest?

5   Wendy is going to visit Japan. She wants to exchange £750 into Japanese yen.

She can go to either her bank or to a travel agent to exchange the money.

Place of exchange	Exchange rate	Commission
Travel agent	£1 = JPY¥168.23	Charge of 2% commission on all sales
Bank	£1 = JPY¥169.89	No commission charged

a   If she goes to the travel agent, how much will she receive?

b   If she goes to her bank, how much will she receive?

c   Where will she get the better deal? Give a reason for your answer.

6   Edgar is planning to visit South Africa. He wants to exchange £600 into South African rand.

He can exchange the money at his bank, at his local travel agent or at the bureau de change at the airport.

Place of exchange	Exchange rate	Commission
Travel agent	£1 = R16.89	Charge of 1.5% commission on all sales
Bank	£1 = R16.71	No commission charged
Bureau de change	£1 = R16.98	Charge of 3% commission on all sales

a If he goes to the travel agent, how much will he receive?

b If he goes to his bank, how much will he receive?

c If he goes to the bureau de change, how much will he receive?

d **i** Where will he get the best deal? Give a reason for your answer.

    ii Which of the three places offers the worst deal? Give a reason for your answer.

7 Sandy is visiting Turkey and Greece. She wants to exchange £300 into euros and £200 into Turkish lira for her trip. Two travel agents offer different deals.

Travel agent	Exchange rates	Commission
Travel agent 1	£1 = €1.22 £1 = ₺3.33	3% commission charged on all sales
Travel agent 2	£1 = €1.20 £1 = ₺3.29	1% commission charged on all sales

a If she uses travel agent 1, how much will she receive of each currency?

b If she uses travel agent 2, how much will she receive of each currency?

c Which travel agent offers the better deal? Give a reason for your answer.

8 John and Amir went on holiday at the same time but to different destinations.

a John travelled to the USA. He exchanged pounds sterling into US dollars and received USD$668. He paid 2% commission of this amount to the travel agent.

    i How much commission did he pay (in pounds sterling)?

    ii How much money did John actually receive?

b Amir travelled to South Africa. He didn't exchange money before he left. Instead, he obtained money in rand using autoteller machines (ATMs). He was charged R25 every time he used an ATM. He used it on four occasions. How much was he charged by the bank (in pound sterling)?

c Who was charged more? Give a reason for your answer.

d What could John or Amir have done to reduce their charges?

9 The table below shows the average monthly exchange rates for pounds sterling to US dollars between January and July 2013.

Month	£1 =
Jan	USD$1.5965
Feb	USD$1.5490
Mar	USD$1.5087
Apr	USD$1.5311
May	USD$1.5291
Jun	USD$1.5466
Jul	USD$1.5169

a **i** If you had exchanged £400 each month, how much in dollars would you have received?

    ii How many more dollars would you have received when the exchange rate was at its highest compared to its lowest?

    iii Express this difference as a percentage of the lowest amount exchanged.

b The table below shows the average monthly exchange rates for pounds sterling to Australian dollars between January and July 2013.

Month	£1 =
Jan	AUD $1.5215
Feb	AUD $1.5014
Mar	AUD $1.4588
Apr	AUD $1.4757
May	AUD $1.5435
Jun	AUD $1.6411
Jul	AUD $1.6574

    i If you had exchanged £400 each month, how much would you have received?

    ii How many more dollars would you have received when the exchange rate was at its highest, compared to its lowest?

    iii Express this difference as a percentage of the lowest amount exchanged.

c By comparing your answers to parts **a iii** and **b iii**, which currency varied more between January and July of 2013?

10 Harold travels to Canada and exchanges £1200 into Canadian dollars.

    a How many Canadian dollars does he receive?

    b He spends CAD$1520. He then travels to the USA and exchanges the remainder of his Canadian dollars into US dollars. How many US dollars does he receive?

11 A travel agency offers the following exchange rates for £1.

Country	Currency	We sell	We buy
Australia	Australian dollar	AUD$1.79	AUD$1.89
Canada	Canadian dollar	CAD$1.74	CAD$1.81
Eurozone countries	euro	€1.21	€1.25
Japan	Japanese yen	JPY¥167.71	JPY¥176.76
Switzerland	Swiss francs	CHF1.48	CHF1.54
USA	American dollar	USD$1.64	USD$1.71

a If you exchanged £350 into Canadian dollars, how many dollars would you receive?

b If you exchanged €450 into pounds sterling, how much would you receive?

c i If you exchanged £250 into each of the currencies in the table above, how much would you would receive in each currency?

    ii If you did not spend the amount exchanged and you took it back to the travel agent, how much money would you receive back?

    iii Which currency does the travel agency make the most money out of when you exchange unused foreign currency back to pounds?

> 'We sell' is the exchange rate used to exchange into foreign currency. 'We buy' is the exchange rate used to exchange back into pounds sterling.

12 The table below shows the exchange rates for pounds sterling to US dollars, euros and Japanese yen for 1 week.

Day	USD$	€	JPY¥
Monday	1.6231	1.1976	164.39
Tuesday	1.6162	1.1899	163.94
Wednesday	1.6201	1.1946	164.14
Thursday	1.6286	1.1997	166.39
Friday	1.6338	1.2041	167.15

a   How many more US dollars would you receive if you changed £500 when the exchange rate was at its highest in comparison to its lowest?

b   How many more Japanese yen would you receive if you changed £500 when the exchange rate was at its highest in comparison to its lowest?

c   How much more would you receive if you changed €500 into pounds sterling when the exchange rate was at its lowest in comparison to its highest?

d   Describe the trend in the exchange rates across this particular week.

e   Which is better? A lower or higher exchange rate if you are:

i   exchanging pounds sterling into a foreign currency

ii   exchanging foreign currency into pounds sterling?

13 Olivia wants to buy a mountain bike. She can buy the bike in the UK for £1599.99. She sees the same bike on sale in France for €1745 or in Norway for NOK17550 (Norwegian krone).

The table shows the exchange rates at the time she wants to buy the bike.

Country	Currency	Exchange rate
France	euro	£1 = €1.21
Norway	Norwegian krone	£1 = NOK10.04

Shipping to the UK from France will cost an extra 10% of the cost in France.

Shipping to the UK from Norway will cost an extra 15% of the cost in Norway.

a   How much will it cost in total to buy the bike and have it shipped to the UK from

i France   ii Norway?

b   Which is the cheapest deal? Give a reason for your answer.

14 Phil is going to Germany and Switzerland on business. He needs to exchange £300 into euros and £400 into Swiss francs. He has five exchange options.

Place of exchange	Exchange rates	Extra charges
Bureau de change at the airport	£1 = €1.17 £1 = CHF1.43	2% Commission
Bank	£1 = €1.16 £1 = CHF1.41	£5 fee
Travel agent	£1 = €1.15 £1 = CHF1.38	£3 fee
Supermarket bureau	£1 = €1.17 £1 = CHF1.42	£4 fee
ATM abroad	£1 = €1.16 £1 = CHF1.41	£1.75 per transaction

a Calculate the amount he would receive in euros and Swiss francs for each option.

b Which place of exchange gives Phil the best deal?

★ 15 Harrison and Sayeeda are going on holiday to Australia to visit family.

They choose to stop in Dubai (United Arab Emirates, UAE) for two nights and Shanghai (China) for three nights before travelling to Brisbane (Australia) for two weeks.

Use the exchange rates shown in the table to answer the questions below.

Country	Currency	Exchange rate
United Arab Emirates	UAE dirham	£1 = AED3.67
China	Chinese yuan renminbi	£1 = CNY¥10.35
Australia	Australian dollars	£1 = AUD$1.79

a They exchange £200 into UAE dirham, £300 into Chinese yuan renminbi and £1800 into Australian dollars. How much of each currency do they receive?

b They spend AED428 in Dubai and CNY¥2220 in Shanghai. They keep AED100 for the stopover on the way home and exchange the remainder of their dirhams and yuan renminbis into Australian dollars. How much do they now have for spending in Australia?

c They spend AUD$3350 in Australia and then spend AED70 at the airport on the way home. They exchange their remaining money back into pounds sterling (GBP) on their return home. How much money do they receive?

16 John and Jamie are travelling by InterRail across Europe. They take £800 between them to pay for accommodation and food. They convert this to euros.

a They travel through France and Germany. They spend €122 on accommodation and €117 on food and drink for five nights. They also buy souvenirs for €19. How much money (in euros) do they now have?

b They then travel to Switzerland where they stay for two nights. They convert €250 into Swiss francs. Their accommodation costs CHF84 and they spend CHF151 on food and drink. How many Swiss francs do they have left?

c Next, they travel to Italy and stay for one night. Hostel accommodation costs them €24 each and they spend €45 on food and drinks. On the way back to the hostel they find that they have lost their InterRail tickets, their ferry tickets and their UK bus tickets for their return home. They find out that it will cost them €116 each to travel to the ferry, the ferry will cost them €64 each and their bus ticket home will cost them £65 each. Using their remaining Swiss francs and euros, do they have enough money to get back home? Give a reason for your answer.

17 Mike is in the duty free shop in Geneva airport (Switzerland). He is buying a designer watch costing CHF490. He has €150 and will pay the rest on his debit card in pounds sterling. How much will be charged to his debit card?

18 Judy is in the duty free shop in Newark Airport (USA) returning home after visiting Canada and the USA. She sees a pair of sunglasses for USD$120. She does not have any US dollars so she hands over CAD$50 and £50. She receives her change in US dollars. How much change does she receive?

19 An art dealer bought a painting in Paris (France) for €550 000 and sold it for USD$730 000 in New York (USA). Did he make a profit or loss? Give a reason for your answer.

GO! Activity

Choose at least two places you would like to visit.

1 Use the internet to research the cost of the following for your chosen locations:

   a the cost of accommodation for 7 nights

   b the cost of a meal (main course and drink) and multiply by 7

   c the cost of at least three activities or places of interest at your chosen location.

2 Use today's exchange rates for the places you would like to visit and work out how much each would cost. Out of the locations you have chosen, which place is the cheapest to visit?

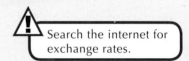

Search the internet for exchange rates.

- I can convert between currencies involving the use of at least three currencies in a multi-stage task.
  ★ Exercise 4A Q2, Q4, Q15

For further assessment opportunities, see the Case Studies for Unit 1 on pages 136–140.

5 Investigating the impact of interest rates on savings and borrowings

This chapter will show you how to:

- investigate the impact of interest rates on loans and credit agreements
- investigate the impact of interest rates on credit cards and store cards
- investigate the impact of interest rates on savings.

You should already know:

- how to investigate the impact of interest rates in a basic situation on loans and credit agreements, credit cards and store cards, and savings
- how to calculate a percentage of a quantity
- how to discuss and illustrate the facts you need to consider when determining what you can afford in order to manage credit and debt, and lead a responsible lifestyle
- how to research, compare and contrast a range of personal finance products and, after making calculations, explain your preferred choices.

Loans and credit agreements

If you take out a **loan**, you are given an amount of money from a bank or finance company. When you do this, you agree to repay the money borrowed, along with interest, in a specified time.

If you take out a loan you will be asked to sign a **credit agreement.** This is a legal contract which outlines the rules and regulations of the loan, including when payments must be made and the amount of interest that must be paid on the loan.

Interest

Interest is an amount added to the loan by the bank or finance company. The bank will calculate the interest for the whole period of the loan.

The loan plus the interest is usually repaid in equal monthly payments which are called **instalments**.

Interest is calculated as a percentage of the loan for each year of the loan agreement. This is sometimes referred to as **per annum** (p.a.) which means 'for each year'.

There are two main types of interest.

- **Simple interest** is calculated as a percentage of the loan, and the amount of interest added is the same each year.

- **Compound interest** is also calculated as a percentage of the loan. However, with compound interest, the interest for the first year is added onto the loan, then the second year's interest is based on the new amount of the loan, and so on. This is called **compounding**. Depending on the agreement, compound interest could be calculated monthly rather than yearly.

> In National 5 Lifeskills Mathematics assume that all <u>loan repayment questions</u> involve calculating **simple interest**.

Example 5.1

Sami is going on a round-the-world trip. To help pay for this holiday, he will borrow £2500.

He is offered three different repayment options at a fixed rate of 9.3% per annum.

Option	Loan term
1	12 months
2	24 months
3	36 months

a Calculate the monthly repayments for each of the three loan options.

b Sami wants to borrow £2500 but can only afford to pay £100 per month. Which option should he choose? Give a reason for your answer.

a   Option 1: Interest = 9.3% of £2500

$$= 0.093 \times 2500$$

> Remember to round all money answers correct to 2 d.p.

$$= £232.50$$

Total amount borrowed = £2732.50

Monthly repayment is £2732.50 ÷ 12 = £227.71

Option 2: Interest = 9.3% of £2500 × 2

$$= 0.093 \times 2500 \times 2$$

$$= £232.50 \times 2$$

$$= £465$$

Total amount borrowed = £2965

Monthly repayment is £2965 ÷ 24 = £123.54

Option 3: Interest = 9.3% of £2500 × 3

$$= 0.093 \times 2500 \times 3$$

$$= £232.50 \times 3$$

$$= £697.50$$

Total amount borrowed = £3197.50

Monthly repayment is £3197.50 ÷ 36 = £88.82

b Sami will need to take the loan out over 36 months because the monthly payments for a 12-month or 24-month loan are more than £100.

Exercise 5A

1   Scott wants to buy a new car which costs £16 505. The car dealership offers three finance agreement options at a fixed rate of 11.2% per annum.

Option	Loan term
1	24 months
2	36 months
3	48 months

   a   i   Calculate the monthly repayments for each of the three loan options.

      ii   How much more does he pay monthly if he takes out the loan over 24 months rather than 36 months?

      iii   If he takes out the loan over 48 months, how much more than the cash price does he pay?

   b   If he is able to pay a deposit of £500, how much will each monthly repayment be if he takes out the loan for 36 months?

2   Two shops sell the same television for the same price of £998.

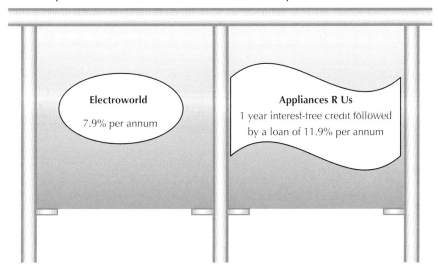

Electroworld

7.9% per annum

Appliances R Us

1 year interest-free credit followed by a loan of 11.9% per annum

By calculating the monthly repayments, which shop offers the better deal if the loan is taken out over:

a   1 year   b   2 years   c   3 years?

> ⚠ 'Interest-free credit' means that you do not pay interest on money borrowed.

★   3   A furniture shop sells a sofa with different special offers for April and May.

April	May
Sofa costs £738	Sofa costs £738 with a discount of £249
Interest-free credit for 3 years	Finance: 12.3% per annum

You decide to buy the sofa and want to pay for it over 3 years.

By calculating the monthly repayments, which month offers the better deal? Give a reason for your answer.

4   The supermarket chain Kings offers financial services through its own bank called Kings Bank. Kings Bank offers loans at different rates.

   a   Calculate the percentage interest per annum if you take out a loan of:

      i   £10000 over 5 years and pay £193.17 per month

      ii   £5000 over 5 years and pay £102.83 per month.

   b   Why do you think the percentage interest is more for a £5000 than a £10000 loan?

5   Rachael needs a new boiler for her flat. The boiler (including labour) will cost £3150. The utility company offers a 3.2% discount if she uses their finance provider, which will lend finance at a rate of 7.9% per annum. She wants to pay the loan over 3 years.

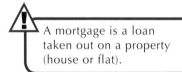 If she uses a different finance provider she is not entitled to the discount offered by the utility company.

   a   How much will Rachael's monthly repayments be if she uses the utility company's finance provider?

   b   If Rachael chooses a bank offering 7.3% per annum, how much will her monthly repayments be?

6   A mortgage company offers a fixed rate of 4.89% per annum for a 25-year mortgage.

   A mortgage is a loan taken out on a property (house or flat).

   Calculate the monthly repayments on a mortgage of:

   a   £125000   b   £423450

7   A kitchen company sells a style of complete kitchen which costs £6498. During January the company offers a discount of 40%. The buyer must pay a deposit of 10% of the discounted cost followed by a finance deal of 15.4% per annum of the remaining cost.

   a   Sophie and Jordan decide to take out the finance deal over 4 years. Calculate their monthly repayments.

   b   If they sign the deal before 14 January they get a discount of 45%. If they bring their own measurements they get £100 off the discounted price. They then pay a deposit of 10% and pay the rest with a finance deal at 15.4% per annum over 4 years.

   How much extra do they save per month if they choose this option?

8   In a particular month, Emily has some unexpected expenses. To help pay for the extra expenses, she takes out a loan of £800 with a payday loan company. The rate of interest is 1834% per annum. If she pays back the loan within one month she is not charged interest. If she does not pay back the loan she must pay back the loan with interest.

   a   If she does not pay back the loan, how much does she need to pay back after 12 months?

   b   If she agrees to pay back the loan over 12 months, calculate her monthly repayments.

   c   i   What advice would you give Emily if she knows that she cannot afford to pay back the loan within one month?

   For more about payday loans, see the Make the Link section near the end of this chapter.

      ii   Why do you think that people take out payday loans even though they have a very high rate of interest?

9   A games console costs £429.99. A shop offers a finance deal with an interest rate of 1.23% per month. If the console is paid for over 18 months, calculate the monthly repayments.

10 Cathie is buying a car which costs £9188. She pays a deposit of £99. She is offered three different repayment options at a fixed rate of 8.2% per annum.

Option	Loan term
1	12 months
2	24 months
3	36 months

    a   Calculate the monthly repayments for each of the three loan options.

    b   Cathie can only make monthly repayments of £300 per month. How much of a deposit, to the nearest pound, will she have to pay on option 3?

11 Sheena and Terry are putting an extension on their house and need to borrow £11 000. They have found three deals.

Deal 1	A building company offers a deal where £510 a month is paid for 2 years.
Deal 2	A bank offers a loan of £11 000 – 5.3% per annum for 2 years.
Deal 3	They go to two different banks to see if a split loan for smaller amounts will be cheaper and get these offers: • bank 1 will charge 5.1% per annum on a loan of £6000 for 2 years • bank 2 will charge 5.6% per annum on a loan of £5000 for 2 years.

Which of these three deals is the cheapest? Give a reason for your answer.

12 Highlands Bank offers deals as shown in the table.

Loan amount (£)	Interest (% p.a.)
1000	18.2
2500	14.15
5000	11.9

    a   Calculate the monthly payment for each loan if repaid over a period of 2 years.

    b   Simone and Andrew want to borrow £5000 to pay for a new central heating system.

        i   If Simone and Andrew each take out a loan for £2500, how much would they each repay per month? How much would they pay combined?

        ii Calculate the difference in monthly repayments between taking out a single loan of £5000 and taking two separate loans of £2500.

        iii Calculate the total difference in monthly repayments between taking out a single loan of £5000 and taking two separate loans of £2500 over a period of 3 years.

Credit cards and store cards

Credit cards

You can get a **credit card** from a bank or finance company which you can use as payment for goods and services. The company which issues your card holds an account for you, and the amount in your account (the amount you owe as a result of using the card to pay for your purchases) is called the **balance**.

You do not need to pay off the balance of your account each month as you have the option of making a smaller payment. However, if you do not pay off the balance in full, interest is added to your account balance, meaning that you have more to pay each month.

The main advantage of a credit card compared to a loan is convenience: it allows small short-term loans, provided the total balance stays below the maximum credit (maximum amount allowed as a balance). Some credit cards even offer rewards and benefits. Other advantages of credit cards are:

- they allow you to be flexible with your money because they do not require fixed monthly payments

- they are a secure way of paying for shopping online

- they enable you to make a claim against the credit card company if goods purchased with the credit card are faulty or fail to arrive, or if the company providing the goods/services collapses or goes bankrupt.

However, credit card companies charge fines if the minimum monthly payment is not met, they will freeze the card so it cannot be used and they allow interest to continue to be added onto the balance. If you do not keep up with repayments, you may find your **credit rating** is negatively affected which means that you will find it harder to get a loan, mortgage or other finance deal in the future.

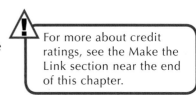
For more about credit ratings, see the Make the Link section near the end of this chapter.

Store cards

A **store card** is a special type of credit card. It is provided by a finance company but you are only able to use it in one particular shop or branch of shops. Shops often have special offers only available to store card holders, and most have a special offer on the first day of its use.

Example 5.2

Hannah wants to buy a desktop computer for £1150.

a Her credit card company charges her simple interest of 20.3% per annum on the balance. Calculate the total amount to be repaid after 1 year (including interest) if she uses her credit card to buy the computer.

b The shop offers her a 10% discount on the price of the computer if she uses their store card. If she uses the store card, she is charged simple interest of 24.3% per annum on the balance. Is this a better deal than the credit card? Use your calculations to justify your answer.

a Interest = 20.3% of £1150

$$= 0.203 \times £1150$$

$$= £233.45$$

Amount to be paid is £1150 + £233.45 = £1383.45

b 10% of £1150 = £115

Price of computer is £1150 − £115 = £1035   ← Work out the discounted price of the computer.

Interest = 24.3% of £1035

$$= 0.243 \times £1035$$

$$= £251.51$$

Amount to be paid is £1035 + £251. 51 = £1286.51

The store card is the better option because she would owe £1286.51 which is lower than £1383.45 which she would owe if she used the credit card.

Exercise 5B

1   Eleanor wants to buy a new dress which costs £65 in a shop.

   a   When she reaches the counter at the shop she sees an advertisement for a store card.

Store card

Get a store card today and receive a £25 discount off today's purchase. Simple interest 32.1% per annum.

       How much will the dress cost her if she pays the balance back a year later?

   b   If she uses her credit card instead, she is charged 1.5% of the cost of the dress by the shop. Her credit card company charges her 18.9% per annum. Calculate the cost of the dress if she is charged interest on the cost and she takes a year to pay back the balance.

   c   Which is the cheaper option, the store card or the credit card?

   d   She returns a month later and buys a pair of trousers for £48 using the store card. If she pays the balance back a year later, how much more will she pay for the trousers using the store card than if she simply paid cash for the trousers?

2   A travel agent offers readers of a magazine a 5% discount off their holidays. Frank and Selma decide to use the offer to go to Bangkok in Thailand. The holiday originally cost £2519.

   They pay for the holiday using their credit card and are charged 2.5% of the cost of the holiday by their credit card company. They are also charged 20.4% per annum for using their credit card. If they pay it back a year later, how much will the holiday have cost in total?

3   A department store is having a pre-Christmas sale and are offering 30% off all purchases. It offers a further 10% off all perfume for store card holders and its store card charges 24.3% simple interest.

   Mark buys these items as presents for his mother and sisters.

Handbag	£42
Pair of boots	£89
Bottle of perfume	£37.50

   What is the balance on his store card a year later?

4   A pharmacy chain offers a reward system of 4 points for every £1 spent at the shop. One point is worth 1p off the price of a new item.

   a   Last year Kelly spent £135 at this pharmacy. How many points has she collected?

   b   Kelly buys these items.

Shampoo	£4.79
Nail varnish	£4.09
Hair straighteners	£89.95

      If she uses the points she has collected as part of her payment, how much does she still need to pay?

    c If she uses her credit card to pay the amount calculated in part **b**, what is the balance on her credit card after a year if she is charged 17.9% simple interest for her credit card?

5 A singer is performing at a large stadium and is charging £65 for each standing ticket and £85 for seated tickets. Brian buys two standing tickets for him and his friend and he buys two seated tickets for his parents.

Brian buys the tickets online from a ticket company. The ticket company charges Brian 11.5% of the cost of each ticket for buying the tickets online. Brian pays by credit card which charges 19.4% simple interest.

What is the balance on his credit card after a year?

6 A shop sells footwear suitable for outdoor occupations, for example, builders, welders, gardeners, outdoor instructors. The shop has a sale one weekend offering a 30% reduction on all footwear.

Ruth is a civil engineer and buys a pair of boots which cost £118.99 (including VAT of 20%) before the sale. She uses the boots for work so she does not need to pay VAT.

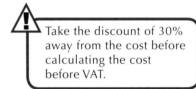
Take the discount of 30% away from the cost before calculating the cost before VAT.

    a How much does she need to pay for the boots if the VAT is removed?

    b She pays for the boots using her store card. If her balance a year later is £82.39, what interest rate does the shop charge for Ruth's card?

★ **7** Sarah buys a football games table for her son for £169.99.

    a Her credit card company charges her simple interest of 19.6% per annum on the balance. Calculate the total amount to be repaid after 1 year if interest is charged.

    b The shop offers Sarah a 12.5% discount on the price of the football table if she uses its store card. It then charges simple interest of 23.4% per annum on the balance. Is this a better deal than the credit card? Use your calculations to justify your answer.

8 Allan buys a bracelet for his wife for £185 at a jewellery shop.

    a His credit card company charges him simple interest. The total amount to be repaid after interest is charged for 1 year is £223.48. What interest rate does the credit card company charge Allan?

    b The jewellery shop offers Allan a 9.5% discount on the price of the bracelet if he uses its store card. It then charges simple interest of 22.9% per annum on the balance. Is this a better deal than the credit card? Use your calculations to justify your answer.

9 Bill buys a tablet computer for £439. His credit card company charges him compound interest of 2.75% monthly. If he pays back £25 at the end of each month, what is his balance after 1 year?

10 June buys a television for £329. Her credit card company charges her 2.85% monthly.

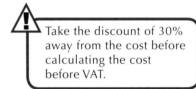
This is known as the annual percentage rate (APR). For more about APR, see the Make the Link section near the end of the chapter.

    a If she does not pay any money back, what is her balance after 1 year?

    b Express the increase as a percentage of the amount of money borrowed.

11 If a company charges 2.18% per month, what is their APR?

12 a Jim wants to take out a loan of £3000 at an interest rate of 3.15% per month.

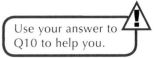

Use your answer to Q10 to help you.

    i How much does he owe if he wants to pay it back after a year?

    ii If he spreads the cost over the year, how much will each monthly repayment be?

  b When he goes to his bank to take out the loan he is told that he can only take out a loan with a rate of interest of 4.35% per month due to his credit rating.

    i How much will he owe if he wants to pay it back after a year?

    ii How much more will he have to pay each month?

Savings

Savings is the money you keep in an account with a bank, building society or other financial institution, usually put aside for some use in the future. Financial institutions offer interest on the balance of savings accounts, and the interest is added to the balance. There are two main types of interest.

- **Simple interest** is calculated as a percentage of the balance, and the amount of interest added is the same each year.

- **Compound interest** is also calculated as a percentage of the balance. However, with compound interest, the interest for the first year is added onto the balance, then the second year's interest is based on the new amount of the balance, and so on.

> In National 5 Lifeskills Mathematics assume that all <u>savings</u> questions involve calculating **compound interest**.

Example 5.3

Chloe wants to put £4000 into a high-interest savings account for 3 years. She has the choice of two accounts at her a bank: one has a fixed rate of interest (3.2%) and another has a **variable** rate of interest.

For more about variable rates of interest see the Make the Link section near the end of the chapter.

a If she chooses the account with the fixed rate and does not take any money out of her account, how much money will be in her account at the end of 3 years?

b The variable rate savings account offers interest of 2.8%, 3.7% and 3.1% for each of the 3 years. If she chooses this account and does not withdraw any money from her account, how much money will be in her account after 3 years?

c Which account would give her the better deal over 3 years?

a There are two methods to answer this question.

Calculate 3.2% of the balance each year and add it on.

Method 1

Year	Interest	Balance
1	3.2% of £4000 is $0.032 \times £4000 = £128$	$£4000 + £128 = £4128$
2	3.2% of £4128 is $0.032 \times £4128 = £132.10$	$£4128 + £132.10 = £4260.10$
3	3.2% of £4260.10 is $0.032 \times £4260.10 = £136.32$	$£4260.10 + £136.32 = £4396.42$

Balance after 3 years is £4396.42.

(continued)

Method 2

Balance is £4000 × 1.032 × 1.032 × 1.032 = £4000 × $(1.032)^3$

$$= £4396.42$$

> One year = 103.2% of previous year. So, multiply £4000 by 1.032 each year for 3 years.

Note: There can be a slight variation in answers to methods 1 and 2. This is because of the rounding involved at each stage of method 1, which can lead to a less accurate answer than method 2.

b

Year	Interest	Balance
1	2.8% of £4000 is 0.028 × £4000 = £112	£4000 + £112 = £4112
2	3.7% of £4112 is 0.037 × £4112 = £152.14	£4112 + 152.14 = £4264.14
3	3.1% of £4264.14 is 0.031 × £4264.14 = £132.19	£4264.14 + £132.19 = £4396.33

Balance after 3 years is £4396.33.

c The fixed rate interest account would be the better deal because £4396.42 is 9p more than £4396.33.

Exercise 5C

1 Calculate the compound interest and the balance on a high-interest account with a fixed rate of interest if you invest:

    a £3500 for 2 years at 3.2% per annum

    b £4200 for 3 years at 2.9% per annum

    c £5400 for 5 years at 3.8% per annum

    d £8000 for 4 years at 4.1% per annum

    e £1200 for 3 years at 2.4% per annum

    f £6950 for 8 years at 3.7% per annum.

2 Calculate the compound interest and balance on a high-interest account with a variable rate of interest if you invest:

    a £3800 for 3 years with interest rates of 2.9%, 3.4% and 3.7%

    b £5600 for 2 years with interest rates of 3.6% and 2.8%

    c £6300 for 4 years with interest rates of 3.2%, 4.1%, 3.9% and 3.5%

    d £1500 for 5 years with interest rates of 3.1%, 4.2%, 3.6%, 3.15% and 2.95%.

3 Simeon puts £4000 into a high-interest account of 11.2% per annum. How many years will it take to double his money?

4 Lindsey puts money into a savings account with a fixed interest rate of 3.4% per annum. At the end of 3 years she has £6036 in her account. How much money did she put into her account 3 years ago?

5 Dwayne puts £4500 into a high-interest account with a fixed interest rate of 3.42% per annum. At the end of each year (after interest has been added), he withdraws £500. Calculate the balance at the end of the fourth year.

★ **6** Anita wants to put £8000 into a high-interest savings account over 3 years.

Her bank offers two types of account:
- fixed rate of interest (3.4%)
- variable rate of interest of 3.1%, 3.7% and 4.1%.

   a If she chooses the fixed rate account and does not take any money out of her account, how much money will be in her account at the end of the 3 years?

   b If she chooses the account offering the variable rate of interest and does not withdraw any money, how much money will be in her account after 3 years?

   c Which account will give her the better deal over the 3 years? Give a reason for your answer.

7 Faisal wants to put £6000 into a high-interest savings account over 4 years.

His bank offers two types of account:
- a fixed rate of interest (3.7%)
- variable rate of interest of 1%, 3.6%, 3.1% and 4.4%.

   a If he chooses the account offering the variable rate and does not take any money out of his account, how much money will be in his account after 4 years?

   b If chooses the fixed rate account and does not withdraw any money, how much money will be in his account at the end of 4 years?

   c Which account will give him the better deal over the 4 years? Give a reason for your answer.

8 Margaret wants to put £8000 into a savings account for 1 year. She has investigated bank interest rates and found the following information.

Bank	Interest rates and other details
Edinburgh Bank	• 3.1% • Bonus of 0.43% for investments over £5000 • Interest taxed at 20%
Lothian Bank	• 3.3% • Interest taxed at 20%
Thistle Bank	• 2.8% • No tax deducted

   a Calculate the interest Margaret would receive from each bank.

   b Advise Margaret which bank she should choose. Use your calculations to justify your answer.

9 Harry puts £3500 into a savings account which has a fixed rate of interest of 1.03% per month.

   a Calculate the compound interest and the balance on this account after 12 months.

   b Express the compound interest as a percentage of the initial amount invested.

> This is known as the annual equivalent rate (AER). For more about AER, see the Make the Link section near the end of the chapter.

GO! Activity

In reality, interest (compound) is added each month on loans, credit cards and bank accounts.

1   To investigate the effect of an interest rate on a loan we can use a spreadsheet or a calculator. For the first part of this question you will explore a loan of £5000 taken out at an interest of 2.5% per month. £200 is paid monthly.

   a   Set up your spreadsheet ready to calculate the interest and balance as shown below.

	A	B	C	D
1	Month	Interest	Balance (£)	
2	0		5000	
3	1			
4	2			
5	3			
6	4			
7	5			
8	6			
9	7			
10	8			
11	9			
12	10			
13	11			
14	12			
15				

   b   Interest is often calculated first before any repayment amount. Calculate how much the interest will be after the first month.

   c   Now add the interest and subtract the £200 repayment. What is the balance after one month?

   d   This is how the first two months should look on a spreadsheet.

	A	B	C
1	Month	Interest	Balance (£)
2	0		5000
3	1	125	4925
4	2	123.125	4848.125
5	3		
6	4		

   Can you format your totals so that they are rounded to the nearest penny?

   e   Can you create a formula to calculate the interest and balance for each month?

   f   How long will it take to repay the loan?

   g   Investigate how long the loan will take to be repaid if:

       i   the interest is increased to 2.9% per month

      ii   the interest is reduced to 2.1% per month

    iii   the monthly payments are increased to £250

     iv   the monthly payments are reduced to £180.

   h   What advice about interest and repayment amounts would you give to a friend taking out a loan?

2   Frankie puts money into a savings account with a variable rate of interest of 2.7%, 3.4% and 3.1% for 3 years. At the end of 3 years she has £7176.66 in her account. How much did she put in her account 3 years ago?

3   Chris is considering three different types of credit card.

- Card A has a monthly interest rate of 1.5%.
- Card B has a monthly interest rate of 2%.
- Card C has a monthly interest rate of 0.75%.

   a   If Chris spends £200 on each card, calculate the balance on each card at the end of 12 months if Chris pays £5 off the balance at the end of each month.

   b   What is the difference in balance between the most and least expensive credit cards?

4   If you take out a payday loan of £600 at a monthly interest rate of 324% and do not pay anything back:

   a   how much will you owe at the end of the 12 months?

   b   what percentage of the amount borrowed will you now owe (APR)?

5   Research interest rates for loans, payday loans, mortgages, credit cards and store cards using a range of sources such as newspapers, magazines and the internet. Choose an amount you would like to borrow and investigate the effects of increasing and reducing the interest rates. Also look at the terms and conditions and consider why some loans are better deals than others.

⚬⚬ Make the Link

Terms and definitions

Payday loans

A payday loan is a loan offered by finance companies. It is taken out in the same way as other loans, but the amount borrowed is due to be paid back in full within a month of borrowing. Companies used to offer loans with no credit check, but new regulations (which came into effect from April 2014) ensure that firms only lend to borrowers who can afford to pay back the loan. The loan is processed on the same day as the application, so the money is usually in the applicant's bank account within 2 hours of their loan application.

Payday loans usually have an extremely high rate of interest which means that the cost of the loan will increase very quickly if the loan is not paid back in full. APR (see next page) is usually at least 1000% for a payday loan.

Taking out a payday loan affects a person's credit rating for 7 years.

(continued)

There has been widespread criticism of short-term loans in the UK. Many people have taken out payday loans as a last resort and have found themselves significantly more in debt as a result.

Credit rating

A credit rating or score is a number which banks and finance companies use to decide whether or not they will lend to an individual person. It is affected by a number of factors: the individual's credit payment history, their current debts, length of time of their credit history, the different types of credit they have used, and how often they have applied for new credit.

When you use credit, you are borrowing money that you promise to pay back within a period of time. A finance lender can use an individual's credit rating as a method of checking the likelihood that the person will pay back the money they have borrowed and any interest accrued.

To improve your credit rating:

- you should make loan payments on time and for the correct amount
- you should avoid having too much credit (for example, more than one credit card plus store cards and loans)
- you should never ignore overdue bills
- you should be aware of what type of credit you have; obtaining credit from payday loan financing companies can negatively affect your score
- you should keep your debt as low as you can
- you should limit your number of credit applications
- you should make sure you have a long history of good credit; a good credit rating is not built overnight, and a longer history is better than a shorter period of good history.

Interest and rates of interest

Interest is paid because the borrower is paying the lender for being allowed to use the lender's money for the period of the loan.

Therefore, if you have borrowed money (in the form of a loan or a credit card), you pay the lender interest.

On the other hand, if you have a bank account which earns interest, it is the bank that pays you interest because it uses your money for other purposes (such as lending).

The **rate of interest** on a loan is a percentage of the amount borrowed which is added to the loan.

A **variable** interest rate on a loan is an interest rate which changes monthly and relies on a rate which is set by the Bank of England (affected by inflation rates).

A **fixed** rate of interest on a loan is an interest rate which doesn't change during the period of the loan. It is usually set at a higher rate of interest than the variable interest rate at the time of taking out the loan, which means that you might end up paying more interest. The advantage of a fixed interest rate is that you have certainly: you can predict future payments and do not have to worry about the interest rate increasing in the future.

Annual percentage rate (APR)

Interest is usually calculated monthly and is calculated as compound interest (see page 62).

APR is the interest rate for the whole year rather than being quoted as the monthly rate.

For example, a credit card might add compound interest monthly at a rate of 2.1%. The balance per month is 102.1% of the previous month which is calculated by multiplying the initial balance by 1.021.

1 year = 1.021^{12} = 1.283 (2.1% added each month)

This means the balance has increased by 28.3%, which is the APR.

APR is used to describe the annual percentage increase on money borrowed.

Annual equivalent rate (AER)

AER is calculated in the same way as APR, but is used to describe the annual percentage increase in savings accounts and current accounts.

Pre-paid credit cards

A pre-paid credit card is a card issued by a bank or finance company with funds preloaded and is used in the same way as a normal credit card. However, instead of buying something using an ordinary credit card (that is, with borrowed money), you are buying something using your own money as the credit is already on the card. It works in a similar way to shop gift cards but can be used in more places and can be used internationally or on the internet.

The major advantage of a pre-paid credit card is that you have the convenience of a credit card but, because you are using your own money, you will not go into debt. The disadvantage is that you must have the funds available in order to pre-pay the card. Many companies offer good online deals, and most are now 'CHIP and PIN' and embossed with your name.

Loyalty card

A **loyalty card** (also known as reward card or points card) is usually a plastic card (similar to a credit or debit card) which has the card holder's name on the front. It shows that the cardholder is a member of a loyalty scheme, usually run by a shop or group of shops. This encourages the card holder to spend in that particular shop, and so build up points which they can either put towards a discount on their current purchase or save points for future purchases.

Payment protection insurance

Credit companies offering loans (including banks) can offer the loan with payment protection or without.

Payment protection insurance (PPI) is an insurance which ensures that the loan will be repaid to the lender if the borrower dies before the loan is completely repaid, or if the borrower loses their job, becomes disabled or ill, or faces any other circumstances that prevents them repaying the loan.

There has been a lot of controversy with PPI as many people in the UK were mis-sold PPI (that is, they were sold PPI packages they did not understand, did not need, or could not use). UK banks have set up multi-billion pound provisions to compensate customers.

PPI has become the most complained-about financial product on record.

- I can investigate the impact of interest rates on loans and credit agreements. ★ Exercise 5A Q3

- I can investigate the impact of interest rates on credit cards and store cards. ★ Exercise 5B Q7

- I can investigate the impact of interest rates on savings. ★ Exercise 5C Q6

For further assessment opportunities, see the Case Studies for Unit 1 on pages 136–140.

6 Using a combination of statistics to investigate risk and its impact on life

This chapter will show you how to:

- develop the link between simple probability and expected frequency
- make use of probability and expected frequency to investigate risk
- investigate risk and its impact on life
- construct and interpret a tree diagram.

You should already know:

- how to convert between fractions, decimal fractions and percentages
- how to calculate the probability of a single event occurring
- how to use probability to make decisions.

Probability

Chance is part of everyday life and the **probability** of an event occurring is the chance that it may happen. Probability is always expressed as a fraction, decimal fraction or percentage.

We use the symbol **P(event)** for the probability that an event will occur.

The probability of an outcome is defined as:

$$P(\text{event}) = \frac{\text{number of ways the event can happen}}{\text{total number of possible outcomes}}$$

The probability of an event has a value between 0 and 1 (including 0 and 1). The chance that an event is **certain** to occur has a value of 1. For example, the probability that the day after Thursday is Friday is 1.

The chance that an event is **impossible** to occur has a value of 0. For example, the probability of winning the National Lottery if you do not buy a ticket is 0.

The probability scale below shows the key words used to describe the probability of an event occurring and their approximate value in the range 0 to 1.

Probability theory is used in many industries, for example in working out how to control the flow of traffic, the running of telephone exchanges, and the study of patterns in the spread of infectious diseases in medicine.

Expected frequency

We can use the probability of an event occurring to predict the **expected frequency** (or value) of the event, that is, the number of times the event will occur in future.

Expected frequency = probability × total number of trials

For example if a **fair** coin is tossed the probability of getting a tail is $\frac{1}{2}$. If the coin is tossed 50 times, the expected frequency of tails is

$$50 \times \tfrac{1}{2} = 25$$

> 'Fair' means all possible outcomes have an equal probability of occurring, that is, there is no **bias**. ⚠️

Relative frequency

The results of an experiment or survey can be used to estimate the probability of a similar event occurring in future. The results are recorded in a frequency table and comparisons can be made using the given data.

Such experimental probability is known as the **relative frequency.**

$$\textbf{Relative frequency} = \frac{\textbf{frequency of the event}}{\textbf{total number of trials}}$$

Relative frequency can be used to predict expected frequency, using data from a previous event. So:

Expected frequency = relative frequency × total number of trials

Example 6.1

a A fair 1 to 6 dice is rolled. What is the probability of rolling a number greater than 4?

b If the dice is rolled 72 times, how many times would you expect a 2?

c Another dice is rolled 84 times. The results are in the table.

Outcome	1	2	3	4	5	6
Frequency	12	20	11	19	9	13

Is this a fair dice? Explain your answer.

a P(> 4) is $\frac{2}{6} = \frac{1}{3}$

> There are 2 numbers greater than 4 (5 and 6) so there are 2 successful outcomes out of 6 possible outcomes. Remember to simplify your fraction.

b P(2) is $\frac{1}{6} \times 72 = 12$

c P(1) is $\frac{1}{6}$

> If the dice is fair, then the probability of each outcome – from 1 to 6 – is the same.

Expected frequency of 1 is $\frac{1}{6} \times 84 = 14$

This is not a fair dice. The numbers 2 and 4 occur much more frequently than would be expected if the dice were fair (20 and 19 times, respectively), and 3 and 5 occur much less frequently than expected if the dice were fair (11 and 9 times, respectively).

Exercise 6A

1   State the probability of each of these events. Choose from the words in the box.

impossible   very unlikely   unlikely   even chance   likely   very likely   certain

   a   Getting a 6 when rolling a 1 to 6 dice.

   b   It will be sunny tomorrow.

   c   The next car you see will be white.

   d   Christmas Day will be on the 25th December this year.

   e   Getting a score of 7 when rolling two 1 to 6 dice.

   f   You will travel back in time.

   g   A person in your class is right-handed.

   h   There are 8 days in a week.

   i   You will live to be 100.

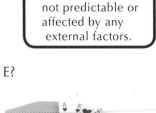

'At random' means that the outcome is not predictable or affected by any external factors.

2   A letter is picked **at random** from the letters in the word STATISTICS.

   What is the probability that the letter will be:

   a   S   **b**   A   **c**   a vowel   **d**   E?

3   A deck of playing cards has four suits: hearts, clubs, diamonds and spades. Each suit has 13 cards: Ace, 2, 3, 4, 5, 6, 7, 8, 9, 10, Jack, Queen and King. A full deck has 52 cards altogether.

   Andrew picks a card from a well shuffled full deck of cards. Find the probability that he picks a card which is:

   a   a Queen   **b**   not a heart   **c**   a black King

   d   red   **e**   an Ace or a 10   **f**   a red 9.

4   The spinner shown is spun.

   a   What is the probability that the spinner will land on a:

   i   2   **ii**   1   **iii**   odd number   **iv**   number less than 4?

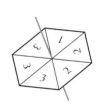

   b   If the spinner is spun 72 times how many times would you expect it to land on a 3?

5   Sharon and Chris are playing a board game using a single 1 to 6 dice. Sharon wins if the number rolled is even. Chris wins if 5 or a 6 is rolled.

   Who has the better chance of winning on the next turn? Use your working to justify your answer.

6 The Accident and Emergency department of a hospital recorded the number of people that were admitted due to alcohol-related incidents over one week.

Age range (years)	0–9	10–19	20–29	30–39	40–49	50–59	60–69	70–79	80+
Number of people admitted to hospital	5	31	37	33	29	18	12	5	2

a What is the relative frequency of a person admitted to hospital being aged 30–39?

b Of the 37 people aged 20–29 years who required treatment, 11 were female. What is the relative frequency of a person admitted to hospital being male aged 20–29 years?

c Give a reason why the 20–29 years age group has the highest number of hospital admissions for alcohol-related incidents.

★ 7 Over a period of time, an Aberdeen hospital carried out several hip replacements for patients suffering from different conditions.

Number of hip replacements	Condition			
	Osteoporosis	Osteoarthritis	Osteogenesis imperfecta	Accidental Injury
Males	275	175	72	37
Females	156	209	142	21

a What is the relative frequency of a male who had a hip replacement?

b What is the relative frequency of a female who had a hip replacement because of an accidental injury?

c What is the relative frequency of a patient who had a hip replacement due to osteoporosis?

d Explain why there are more males who need a hip replacement because of accidental injury than females.

8 The government recommends that anyone over the age of 11 should consume 5 portions of fruit and vegetables every day. In a study carried out by the Food Standards Agency in 2012 it was discovered that, on average:

- adults over the age of 18 were consuming on average 4 portions a day with only 30% of all adults reaching the 5-a-day recommendation

- 11–18 year olds were consuming on average 3 pieces of fruit and vegetables a day with only 7% of all 11–18 year olds reaching the 5-a-day recommendation.

a In 2012 there were 490 000 11–18 year olds and 4 103 000 adults over 18 living in Scotland. How many people over the age of 11 were reaching the 5-a-day recommendation?

b i Approximately how many portions of fruit and vegetables should an 11–18-year-old consume over a 365-day year?

   ii Approximately how many portions of fruit and vegetables did the average 11–18-year-old consume over a 365-day year?

c A lack of healthy diet can lead to obesity or type 2 diabetes in later life.

   What is the expected frequency of 11–18 year olds:

   i becoming obese, if the probability that a person becomes obese or morbidly obese is 14.4%

   ii developing type 2 diabetes, if the probability of a person developing type 2 diabetes by the age of 65 is 0.05?

d What recommendations would you give to the government to help people consume their recommended intake and how do you think this could be implemented?

9 Hayley has a new job selling new and used cars. She records the number of cars she sells in her first 6 months in the job.

Month	June	July	August	September	October	November
New cars sold	1	2	4	6	4	6
Used cars sold	5	4	7	8	7	3

a What is the relative frequency of selling:

   i new cars   **ii** used cars?

b If she sells 8 cars in December, what is the expected frequency of selling:

   i a new car   **ii** a used car?

c Why do you think she might sell fewer cars in December than in October and November?

10 In the 2014 Commonwealth Games held in Glasgow, 261 gold, 261 silver and 302 bronze medals were awarded. The table shows the number of medals won by Team Scotland at the Games.

Type of medal	Gold	Silver	Bronze	Total
Number won	19	15	19	53

a What is the relative frequency of Team Scotland winning:

   i a gold medal

  ii a silver medal

 iii a bronze medal?

b In the Swimming Championships 44 gold, 44 silver and 45 bronze medals were awarded.

Based on your answers to part **a**, calculate the expected frequency of Team Scotland winning these medals in the Swimming Championships:

   i a gold medal

  ii a silver medal

 iii a bronze medal.

c The table shows the actual numbers of swimming medals won by Team Scotland.

Type of medal	Gold	Silver	Bronze	Total
Number won	3	3	4	10

Using your answers to part **b**, comment on the performance of Team Scotland in the Swimming Championships compared to their performance in the rest of the Games competition.

11 In 2000, the world's population was estimated to be 6.1 billion.

According to the World Health Organisation (WHO), in 2000, 175 million cases of malaria and 682 000 deaths from malaria were recorded.

   a In 2000, what was the relative frequency of someone in the world:

      i contracting malaria   **ii** dying from malaria?

   b In 2013, the world's population was estimated to be 7.2 billion.

     Based on your answers to part **a**, what was the expected frequency in 2013 of someone in the world:

      i contracting malaria   **ii** dying from malaria?

   c In 2013, 207 million cases of malaria and 627 000 deaths from malaria were recorded.

     Compare these figures to your answers in **b**. Comment on the results and give a reason why this might be the case.

> ⚠ One billion is one thousand million (1 000 000 000 or 10^9). In the UK, one billion used to mean one million million, but the US definition of billion (one thousand million) has generally taken over, and 'one trillion' is used to mean one million million.

Combined events

Two events are **independent** if the probability of each occurring does not depend on the other. Independent events can occur at the same time.

The probability of two **combined** events, A and B, occurring at the same time is given as:

   P (A and B) = P(A) × P(B)

Combined events can be represented using **sample space diagrams** and **tree diagrams**.

Tree diagrams show the probability of getting specific results in a given sequence of events.

Example 6.2

Two 1–6 dice, one blue and one red, are rolled at the same time.

The **sample space diagram** shows all the possible outcomes.

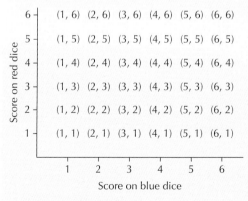

> Each dice has the numbers 1, 2, 3, 4, 5 and 6 on its faces, so there are 6 × 6 = 36 possible combinations.

   a What is the probability of getting a 4 on the blue dice and a 2 on the red dice?

   b What is the probability of getting a total score of 5?

(continued)

a $P(4, 2) = \frac{1}{36}$

$P(4, 2)$ is $\frac{1}{6} \times \frac{1}{6} = \frac{1}{36}$

> Write the combined event as $P(4, 2)$. You can see the answer from the sample space diagram.

> Or, you can calculate the result using the formula. Assuming equally likely outcomes, the probability of rolling any face is $\frac{1}{6}$.

b There are 4 possible ways of getting a total score of 5: (4, 1), (3, 2), (2, 3) and (1, 4).

So the probability of getting a total score of 5 is $P(5) = \frac{4}{36} = \frac{1}{9}$

> You can see the scores from the sample space diagram.

Example 6.3

A pack of six cards has 3 cards with blue squares, 2 cards with red triangles and 1 card with a green circle.

Two cards are picked at random from the pack and the results noted but the first card is replaced before the second card is taken.

a Draw a tree diagram to represent this information.

b What is the probability of picking a blue square followed by a red triangle?

c What is the probability of picking 2 red triangles?

d What is the probability of picking a blue square and a green circle, in any order?

a First pick:

$P(\text{square}) = \frac{3}{6}$ $P(\text{triangle}) = \frac{2}{6}$ $P(\text{circle}) = \frac{1}{6}$

> When picking the first card there are three possible outcomes.

> Each branch represents each possible outcome.

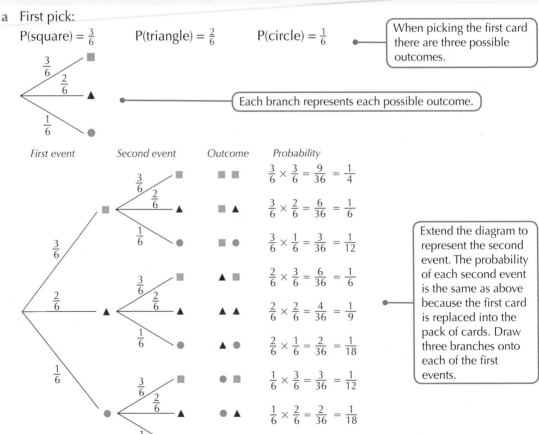

> Extend the diagram to represent the second event. The probability of each second event is the same as above because the first card is replaced into the pack of cards. Draw three branches onto each of the first events.

b P(square, triangle) = $\frac{3}{6} \times \frac{2}{6} = \frac{6}{36} = \frac{1}{6}$

c P(two red triangles) = $\frac{2}{6} \times \frac{2}{6} = \frac{4}{36} = \frac{1}{9}$

d P(blue square, green circle) = $\frac{3}{6} \times \frac{1}{6} = \frac{3}{36} = \frac{1}{12}$

P(green circle, blue square) = $\frac{1}{6} \times \frac{3}{6} = \frac{3}{36} = \frac{1}{12}$

Therefore P(triangle and circle) = $\frac{1}{12} + \frac{1}{12} = \frac{2}{12} = \frac{1}{6}$

> The probability of each outcome can be calculated by selecting the outcome, following the probability on each branch and multiplying together. Remember to write fraction answers in their simplest form whenever possible.

> Calculate the probability of a blue square followed by a green circle, and a green circle followed by a blue square, then add the probabilities together.

Exercise 6B

1 Three coins are tossed at the same time.

   a Copy and complete the tree diagram to show all the possible outcomes.

First event   Second event   Third event   Outcome   Probability

(H, H, H) $\frac{1}{2} \times \frac{1}{2} \times \frac{1}{2} = \frac{1}{8}$

   b Use the tree diagram to work out the probability that, if a coin is tossed three times, the outcome is:

     i three tails

     ii a head and two tails (in any order)

     iii at least one head.

   c Express your answers to part **b** as a decimal fraction.

   d Express your answers to part **b** as percentages.

2 A bag contains 12 black and 15 white counters. A counter is taken out at random, then replaced and a second counter is taken.

   a What is the probability that the first counter is black?

   b What is the probability that the second counter is white?

   c Draw a tree diagram to show the probability of each pick being a black or a white counter.

   d What is the probability that:

     i both counters are black

     ii one counter is black and the other is white (in any order)?

★ 3 Samantha buys her groceries at her local shop, which is staffed by Rita and Grant. She
often shops in the morning then the same afternoon.

She has calculated that the probability of Grant serving her in the morning is $\frac{1}{3}$ and in the
afternoon is $\frac{1}{2}$.

a Copy and complete the tree diagram.

| Morning | Afternoon | Outcome | Probability |

$$\frac{1}{3} \diagup G \diagup \frac{1}{2} G \quad (G, G)$$
$$R$$
$$R \diagup G$$
$$R$$

b Use the diagram to calculate the probability that:

   i Grant serves Samantha in both the morning and afternoon

   ii Samantha is served by Rita in the morning and Grant in the afternoon.

c If Samantha visits the shop twice daily on 12 occasions, on how many days would
she expect to only be served by Rita?

d Grant increases his hours in the shop so the probability of him serving Samantha in
the mornings is now $\frac{1}{2}$.

What is the new probability that Grant serves Samantha in both the morning and
afternoon?

4 Jane is a medical sales rep. The probability that she is working in Edinburgh on a Monday
is $\frac{1}{5}$ and the probability that she is working in Edinburgh on a Tuesday is $\frac{2}{9}$.

What is the probability that she is:

a working in Edinburgh on both days

b working in Edinburgh on a Monday but not on a
Tuesday

c not working in Edinburgh on either day?

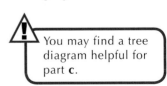
Use a tree diagram or
use the formula P(A
and B) = P(A) × P(B).

5 In a college for performing arts many students study more than one discipline. The
probability that a student chosen at random studies drama is 0.8, the probability that a
student takes dance is 0.3 and the probability that a student takes singing lessons is 0.4.

What is the probability a student selected at random

a takes singing, drama and dance

b does not take drama, singing lessons or dance

c participates in two of the three disciplines mentioned
above?

You may find a tree
diagram helpful for
part **c**.

6 Hamza and Matthew are going to take a swimming test. The probability that Hamza will pass the swimming test is 0.85. The probability that Matthew will pass the swimming test is 0.6. Whether each passes or fails is independent of the other passing or failing.

   a   Copy and complete the probability tree diagram.

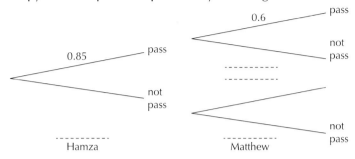

   b   What is the probability that both Hamza and Matthew pass the swimming test?

   c   What is the probability that one of them passes the swimming test but the other does not?

7 Thomas has to take a three-part French exam.
   • He has a 0.4 chance of passing the speaking part.
   • He has a 0.5 chance of passing the listening part.
   • He has a 0.7 chance of passing the writing part.

   a   Draw a tree diagram to represent this information.

   b   If he passes all three parts, his father will give him £20. What is the probability he gets the money?

   c   If he passes two parts, he can re-sit the third part. What is the probability he will have to re-sit?

   d   If he fails all three parts, he will have to withdraw from the course. What is the probability he will have to withdraw?

8 The driving test is in two parts: a written test and practical test.
   • 90% of people who take the written test pass.
   • 60% of people who take the practical test pass.

   a   What is the probability that someone passes the written test and practical test at their first attempt?

   b   What is the probability that someone passes the written test but fails the practical test at their first attempt?

9 In Britain four people in ten go abroad for their holiday.

   Two people are interviewed at random about where they went for their last holiday.

   a   Copy and complete the tree diagram to show the possible outcomes.

   b   What is the probability that the two people interviewed both went abroad for their holiday?

   c   If another two people are interviewed, what is the probability that one of them went abroad for their holiday and the other had their holiday in Britain?

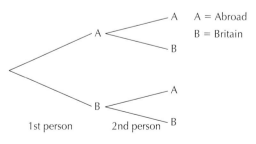

Risk and probability

Risk is the potential that a chosen activity or action may result in a loss or undesired outcome, and can be described as the probability of uncertain future events. Risks are taken every day in various activities such as sport or other leisure activities, driving a car, cooking, investing money, and so on. Predictions of risk are determined by known factors or statistics of past events.

Managing risk is an important factor in areas such as health and safety, insurance, finance and security.

Example 6.4

In 2011 the Department for Transport recorded 146 192 road accidents in the UK. The accidents were recorded by daylight/darkness and associated weather conditions. The results are shown in the table.

Number of road accidents, 2011

Weather conditions	Time of day	
	Daylight	Dark
fine	97 385	29 895
rain	10 687	7 078
snow	313	179
fog	315	340

a   Using the data in the table, what is the relative frequency of an accident occurring:

   i   during daylight when it is snowing

   ii   in the dark when the weather is fine?

b   There were 195 723 accidents recorded in 2013. Based on the 2011 statistics, how many of these would you expect to have happened during daylight when there is fog?

c   i   Which time of day and conditions give the highest relative frequency of an accident? What is this relative frequency?

   ii   Do you think that this condition is the riskiest driving condition? What other factors would you have to consider?

a   i   P(daylight and snowing) $= \frac{313}{146\,192} = 0.00214$ (3 s.f.)

   ii   P(dark and fine) $= \frac{29\,895}{146\,192} = 0.204$ (3 s.f.)

> It is not essential to change this to a decimal fraction, but it helps to use decimal fractions or percentages when comparing probability and relative frequency.

b   Expected frequency $= \frac{315}{146\,192} \times 195\,723 = 422$

> Round your answer to the nearest whole number.

c   i   The highest relative frequency of an accident occurring is during daylight when the weather is fine.

   ii   This is not the riskiest combination of factors, because it does not take into account the actual numbers of people driving in each set of conditions: more people drive in daylight and when conditions are considered fine than at other times.

Exercise 6C

1 According to statistics from the Department of Transport, there were 1714 reported road deaths in Great Britain in 2013. The table shows their data.

Death type	Number of fatalities
Car occupant	792
Pedestrian	398
Motorcyclist	329
Pedal cyclist	107
Other	88
Total	**1714**

a What is the relative frequency of a pedestrian road death?

b What is the relative frequency of a reported road death being a motorcyclist or pedal cyclist?

c There were 2531 road deaths reported in 2008. How many of them would you expect to have been car occupants?

d i Why do you think that the number of road deaths has decreased since 2008?

ii State three recommendations that could help reduce the number of car occupant deaths, particularly with young or inexperienced drivers.

2 The table shows the number of people assisted by the search and rescue services around the coast of the UK.

Type of accident	Number
Beach incidents	675
Diving	68
Water sports	255
Fishing vessel	490
Commercial vessel	124

a What is the probability that a person assisted by the search and rescue services is a diver?

b What is the probability that assistance required from the rescue services is for a vessel?

c Explain why there are more beach accidents than any other type of accident.

3   The table shows the fire statistics recorded by the Scottish Government for 2012–13.

Type of fire	Number attended by firefighters
Primary (buildings, vehicles and fires involving casualties)	11177
Secondary (outdoor fires unless casualties involved)	14105
Chimney (chimney structure fire unless casualties involved)	1331
Total	26613

a   What is the relative frequency of a fire (requiring the fire service to be called out) being a secondary fire?

b   There were 46 fatalities in primary fires in 2012–13. What was the relative frequency of survival if someone was in a building where a primary fire took place?

c   Explain why chimney fires have a lower relative frequency than other fires.

d   What can the Fire Service do to reduce the risk of a fire starting:

   i   in buildings

   ii   outdoors?

4   A company is comparing the success rate of different medication trials and records these results.

Medication trial	Number of successful trials	Number of unsuccessful trials
Liver cancer	57	32
Pancreatic cancer	24	62
Alzheimer's disease	116	47
Heart disease	81	33
Diabetes	9	56

a   What is the relative frequency of a successful medication trial for liver and pancreatic cancer?

b   What is the relative frequency of a successful medication trial?

c   Explain why some patients are given a placebo when different medicines are trialled.

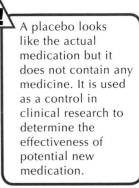

A placebo looks like the actual medication but it does not contain any medicine. It is used as a control in clinical research to determine the effectiveness of potential new medication.

★ 5 A pharmaceutical company has developed three different antibiotics to treat acne.
They performed a large-scale trial with all three antibiotics to test for side effects.

The table shows the number of people tested and whether or not they experienced side effects.

Antibiotic	Side effects	No side effects
Drug 1	1632	7564
Drug 2	1425	6734
Drug 3	1245	6139

a What is the relative frequency of a person not experiencing side effects if they used:

  i Drug 1

  ii Drug 2

  iii Drug 3?

> More information is needed as there is no indication of the severity of side effects.

b What is the relative frequency of experiencing a side effect from any of the drugs?

 c Based on your answer to part a, which antibiotic would you recommend?

 d Are there other factors you need to consider?

★ 6 The Health and Safety Executive (HSE) aim to reduce the number of injuries and fatalities in the workplace in the UK.

Their results in 2011–12 revealed that the relative frequency of a fatal accident in the workplace in the UK was 0.000 005 5.

a If the working population of the UK was approximately 33 million, estimate how many people had a work-related fatal accident in 2011–12.

b There were 148 fatal injuries in 2012–13.

  i What is the relative frequency of fatal injuries in 2012–13?

 ii Has the HSE achieved its aim? Give a reason for your answer.

c The working population of Scotland in 2012–13 was approximately 2 450 000.

  i Using your answer to part b i as the probability of a fatal accident, estimate the number of workplace fatalities expected in Scotland in 2012–13.

 ii There were 22 work-related fatalities in Scotland in 2012–13. Is this an improvement on the statistics for the rest of the UK? Give a reason for your answer.

d What can the government do to reduce fatality statistics?

★ 7   According to the American College of Emergency Physicians, approximately 21 500 gymnasts aged 6–17 years were injured in 2005 in the USA.

The table shows the relative frequency of injuries according to location in body.

Part of body	Relative frequency
Upper body	0.42
Lower body	0.34
Head/neck	0.13
Torso	0.1

a   Approximately how many injuries were:

   i   upper body   ii   head/neck?

b   If 82% of those injured were female, how many females were injured?

c   If there are 5.2 million 6–17 year olds participating in gymnastics, what is the relative frequency of a gymnast being injured?

d   In 2005 in the USA, approximately 775 000 6–17 year olds were injured out of 30 million participants in all sporting events.

   i   What is the relative frequency of a person being injured in sport?

   ii   Is gymnastics a higher or lower risk than other sporting activities? Give a reason for your answer. Use your answers to parts **c** and **d i** to help you.

e   What could sporting coaches do to minimise the risk involved in gymnastics?

Risk assessment

Risk assessment is the process of analysing an activity or event to forecast potential risks and to then take steps to reduce possible exposure to risk. In most cases in everyday life, the probability of a risk is not significant. The second element of a risk assessment is a consideration of the severity of the potential hazard.

For example, if you are comparing travel by road or by air, the probability of a road accident is higher but the severity of the hazard in travelling by air is much greater.

A scale of risk is shown in the diagram below, showing the probability (calculated between 0 and 1) and the severity of the hazard (calculated between 0 and 100). (0 represents 'No hazard' and 100 represents 'Severe hazard'.)

In schools this process has been assisted through the advice from an organisation called SSERC. They provide risk assessments for popular scientific experiments. For other activities they have given the following advice:

- look for the hazards

- decide who might be harmed and how

- evaluate the risk and decide whether the existing precautions are adequate or whether more needs to be done

- record your findings

- review your assessment and revise it as necessary.

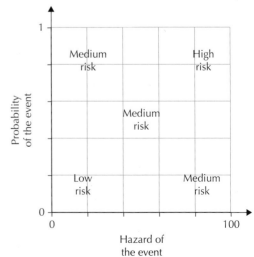

Example 6.5

The grid shows the risk factors of five events: A, B, C, D, E. Associate the five activities with each event on the grid:

1 a person driving very fast on a winding road having an accident

2 a person carrying a mobile phone being struck by lightning

3 a person on a train sitting next to someone with a cold catching a cold as well

4 a person burning their fingers while removing a hot tray from the oven

5 a milk tanker spilling its load on a road.

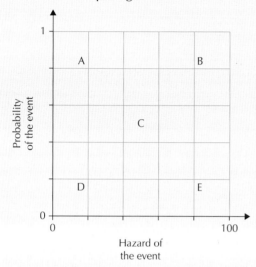

Activity 1 – Outcome B: high probability and high risk. There is a high chance of someone having an accident if they drive at speed on a winding road and if they do, they are likely to die.

Activity 2 – Outcome E: low probability and high hazard. The likelihood of being struck by lightning is very low but the hazard is very serious and may result in death.

Activity 3 – Outcome A: high probability and low risk. They are very likely to catch a cold but the hazard is not serious.

Activity 4 – Outcome C: medium probability and medium risk. It is likely a person will hurt their hand and it is likely to be quite painful but not life threatening.

Activity 5 – Outcome D: low probability and low hazard. It is uncommon for a milk tanker to spill its load, and, if it does happen, it is a low hazard to those involved.

Exercise 6D

★ 1   Make a copy of the probability-hazard diagram.

Think of different activities (similar to those above) and their associated likelihood and hazards.

Write down at least two examples of your own for each type of activity and mark them on your probability-hazard diagram:

A: high probability and low risk

B: high probability and high risk

C: medium probability and medium risk

D: low probability and low hazard

E: low probability and high hazard

Compare these with others in your class.

2   Complete the risk assessment sheet below for planning a school trip to a science museum. Include a journey by bus to and from the museum, an afternoon at the science museum, trips and falls, supervision of pupils.

Use the following ratings:

- Severity and Likelihood on a scale of 1–4 with 1 representing low and 4 representing high.

- Risk rating is L = low, M = medium and H = high.

What has the potential to cause harm (hazards)	Who and how many people might be at risk?	What measures are already in place?	Severity	Likelihood	Risk rating

3   Repeat the process in Question 2 for the planning of these sports activities involving teams:
   a   a football match
   b   a netball match
   c   a sports day.

4   Repeat the process in Question 2 for an overnight Duke of Edinburgh expedition.

- I can develop the link between simple probability and expected frequency. ★ Exercise 6A Q7

- I can construct and interpret a tree diagram. ★ Exercise 6B Q3

- I can make use of probability and expected frequency to investigate risk. ★ Exercise 6C Q5

- I can investigate risk and its impact on life. ★ Exercise 6A Q7 ★ Exercise 6C Q6 ★ Exercise 6D Q1

For further assessment opportunities, see the Case Studies for Unit 1 on pages 136–140.

7 Using statistics to analyse and compare data sets

This chapter will show you how to:

- analyse and compare two or more data sets using mean and standard deviation
- analyse and compare two or more data sets using median, interquartile range and semi-interquartile range
- construct and interpret a boxplot.

You should already know:

- how to calculate the measures of central tendency (mean, median and mode)
- how to calculate and interpret the range (measure of spread).

Standard deviation

Standard deviation is a measure of the spread of data using all the data in the sample. It measures the deviation of each value from the **mean** and it shows how much the data in the sample varies from the mean average.

> Mean = sum of all the values ÷ number of values

Standard deviation is used to describe the spread of data within a data set. A higher standard deviation shows that the numbers are more spread out; that is, there is greater variation between numbers than in a data set with a lower standard deviation. A higher standard deviation could also suggest that the numbers in the data set are less consistent than in a data set with a lower standard deviation.

There are two methods for calculating the standard deviation.

Method 1
Method 1 uses the following formula:

$$\textbf{standard deviation} = \sqrt{\frac{\Sigma(x - \bar{x})^2}{n - 1}}$$

The table shows the meaning of the symbols in the formula.

Symbol	Meaning
Σ	sum of
x	each value
$\bar{x}$	mean value (pronounced 'x bar')
n	number of values

So:

$x - \bar{x}$ represents each deviation

and

$\Sigma(x - \bar{x})^2$ represents the total of the squared deviations.

Method 2

Method 2 uses the formula:

$$\text{standard deviation} = \sqrt{\frac{\sum x^2 - \frac{(\sum x)^2}{n}}{n-1}}$$

The table shows the meaning of the symbols in the formula.

Symbol	Meaning
$\sum x^2$	square each value then add together
$(\sum x)^2$	add all the values together then square the total

This is a rearranged form of the formula in method 1 and is more useful if:

- you are not required to calculate the mean

- you are asked to calculate the standard deviation of a large sample of numbers

- you want to reduce rounding errors.

Both formulae for standard deviation are given in the final National 5 Lifeskills exam.

Example 7.1

During a physics experiment 12 students were each given a dry Leclanché cell (an early form of battery).

Each cell gave continuous light for a period of time, which the students recorded (in minutes). The results were:

$$110, 127, 132, 122, 141, 135, 119, 125, 124, 119, 128, 118$$

Find the mean and standard deviation of the length of time the cells gave out continuous light.

Method 1

$$\text{Mean } (\bar{x}) = \frac{110 + 127 + 132 + 122 + 141 + 135 + 119 + 125 + 124 + 119 + 128 + 118}{12}$$

$$= \frac{1500}{12}$$

$$= 125 \text{ minutes}$$

x	$x - \bar{x}$	$(x - \bar{x})^2$
110	−15	225
127	2	4
132	7	49
122	−3	9
141	16	256
135	10	100
119	−6	36

x	$x - \bar{x}$	$(x - \bar{x})^2$
125	0	0
124	−1	1
119	−6	36
128	3	9
118	−7	49
$\sum x = 1500$		$\sum(x - \bar{x})^2 = 774$

$$\text{Standard deviation} = \sqrt{\frac{\Sigma(x - \bar{x})^2}{n - 1}}$$

$$= \sqrt{\frac{774}{11}}$$

$$= 8.3883$$

$$= 8.4 \ (1 \ \text{d.p.})$$

Method 2

x	x^2
110	12 100
127	16 129
132	17 424
122	14 884
141	19 881
135	18 225
119	14 161
125	15 625
124	15 376
119	14 161
128	16 384
118	13 924
$\Sigma x = 1500$	$\Sigma x^2 = 188\,274$

$$\text{Standard deviation} = \sqrt{\frac{\Sigma x^2 - \frac{(\Sigma x)^2}{n}}{n - 1}}$$

$$= \sqrt{\frac{188274 - \frac{1500^2}{12}}{11}}$$

$$= 8.3883$$

$$= 8.4 \ (1 \ \text{d.p.})$$

Comparing mean and standard deviation

The smaller the standard deviation, the closer the values are to the mean, which means that the values are more consistent.

The larger the standard deviation, the more spread out the values are from the mean, which means that the values are less consistent.

Use this information to help you to answer comparison questions.

Example 7.2

The students in Example 7.1 also performed an experiment on 12 modern AA batteries and recorded the length of time each battery was able to light a torch continuously.

They discovered that the mean time was 150 minutes and the standard deviation was 10.6 minutes.

Make two valid comparisons between the dry Leclanché cell and the modern AA batteries.

- On average the modern AA batteries lasted longer than the Leclanché cells.

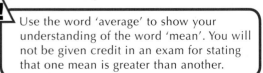 Use the word 'average' to show your understanding of the word 'mean'. You will not be given credit in an exam for stating that one mean is greater than another.

- The life of the AA batteries was less consistent than the Leclanché cells.

 Using the words 'consistent' or 'spread' is acceptable here. Do not use the word 'range' as this has a different meaning.

Exercise 7A

1   Calculate the mean and standard deviation of:

   a   5, 7, 7, 9, 12

   b   15, 16, 18, 21, 22, 22

   c   8, 10, 10, 12, 15

   d   30, 21, 26, 42, 44, 44

2   Calculate the mean and standard deviation of these data sets.

   a   The number of points scored by a rugby team in six consecutive games:

   19   27   11   35   31   21

   b   The response time (in minutes) of seven callouts in one evening from a fire station:

   2   20   31   45   9   31   16

   c   The cost of a medium cup of Americano coffee in five different coffee shops:

   £1.90   £2.05   £1.70   £1.95   £2.10

3   The cost of an adult ticket at a popular concert venue for six different concerts is shown below:

   £22   £35   £28   £42   £26   £45

   a   i   Calculate the mean price of a ticket.

   ii   Calculate the standard deviation of the price of a ticket.

   b   If the tickets are bought online, a booking fee of £1.50 is added to the price of each ticket. Without calculating, write down:

   i   the mean price of the six ticket prices for online bookings

   ii   the standard deviation of the ticket prices for online bookings.

4   The annual salaries of five employees in a small firm are:

   £18 000   £24 000   £26 000   £27 000   £45 000

   a   i   Calculate the mean salary.

      ii   Calculate the standard deviation of the salaries.

   b   Each employee is given a £2000 pay rise. Write down:

      i   the new mean salary

      ii   the new standard deviation.

★ 5   a   The table shows the lowest daily temperature (°C) for a week in December in Inverbervie.

Day	Sun	Mon	Tues	Wed	Thurs	Fri	Sat
Temperature (°C)	−3	0	1	−1	4	2	−3

   Calculate the mean and standard deviation for the temperature in Inverbervie.

   b   This table shows the lowest temperatures for the same week in London.

Day	Sun	Mon	Tues	Wed	Thurs	Fri	Sat
Temperature (°C)	2	3	0	1	−2	7	3

   Calculate the mean and standard deviation for the temperature in London.

   c   Make two valid comparisons between the temperatures in Inverbervie and London during the week in December.

6   Eight women were asked how many pairs of shoes they owned. Their answers were:

   22 30 15 45 26 35 12 31

   a   Calculate the mean and standard deviation of the number of pairs of shoes owned by this sample of women.

   b   Eight men were also asked how many pairs of shoes they owned. Their answers were:

      5 9 13 9 4 8 10 14

   c   Calculate the mean and standard deviation of the number of pairs of shoes owned by this sample of men.

   d   Write down a valid comparison between the group of men and women in this survey.

7   a   A fast food shop provides a drive-in service for customers. They record the length of time (in minutes) that customers wait before being served during peak times in the week. Their results for six customers were:

      5 7 10 8 11 13

      Calculate the mean and standard deviation of the waiting time.

   b   A doughnut shop also provides a drive-in service for customers. They record the length of time (in minutes) that customers wait before being served during peak times in the week. Their results for six customers were:

      12 15 23 45 7 30

      Calculate the mean and standard deviation of the waiting time.

   c   Make two valid comparisons between the waiting times in the fast food shop and the doughnut shop.

8 **a** The weights of five giant male pandas are:

124 kg   105 kg   146 kg   101 kg   114 kg

Calculate the mean and standard deviation of the weight of this sample of giant male pandas.

b The weights of five giant female pandas are:

95 kg   103 kg   107 kg   92 kg   98 kg

Calculate the mean and standard deviation of the weight of this sample of giant female pandas.

c Make two valid comparisons between the weights of male and female giant pandas.

d Edinburgh Zoo keeps two giant pandas. Yang Guang is a giant male panda and weighs 155 kg. Tian Tian is a female giant panda and weighs 101 kg.

Comment on the weight of Yang Guang and Tian Tian compared to the mean weight of male and female giant pandas calculated in parts **a** and **b**.

9 A lot of rain fell in the south of England in December 2013. Daily rainfall for one week was recorded at two weather stations; one in Dorset and one in East Sussex.

a The daily rainfall (in mm) recorded at Dorset for the week was:

17 9 12 37 10 21 13

Calculate the mean and standard deviation of the rainfall in Dorset.

b The daily rainfall (in mm) recorded at East Sussex for the week was:

5 12 13 23 13 11 7

Calculate the mean and standard deviation of the rainfall in East Sussex.

c Make two valid comparisons between the rainfall in Dorset and East Sussex.

10 A company claims that there are 100 drawing pins in each box.

a Find the mean and standard deviation of this sample of boxes correct to 1 decimal place:

101 102 99 97 98 102 101

b The Advertising Standards Agency states that, for the company's claim to be valid, that the mean of a sample must be between 99 and 101, and the standard deviation of a sample should be less than 2.

Based on this sample, is the company's claim valid?

11 **a** Gordon records these scores for his most recent rounds of golf:

77 78 74 73 80 74

Calculate the mean and standard deviation of Gordon's golf scores.

b Gordon's friend Davey plays at the same golf course. These are his scores for his last six rounds:

80 72 78 83 71 72

Calculate the mean and standard deviation of Davey's golf scores.

c Make two valid comparisons between Gordon and Davey's scores.

Interquartile range

There are potential disadvantages of using mean and standard deviation.

- The mean and standard deviation use all the data in a data set. This means that an outlying piece of data will give an average and measure of spread which do not reflect the rest of the data.

- You need a calculator or spreadsheet to work out the mean and standard deviation.

The **interquartile range** is a different measure of the spread of data which uses the **median** of the data. It only takes into account the middle 50% of the data so it is not affected by outlying pieces of data and is easier to calculate than standard deviation, especially when large data sets are involved.

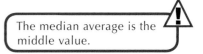

The median average is the middle value.

- To find the median, first list the numbers in numerical order (usually smallest to largest).

- If the list has an odd number of values, the median will be the middle value.

- If the list has an even number of values, the median will be the mean of the two middle values.

Quartiles

Before you can calculate the interquartile range, first you need to calculate the **quartiles** of the data. Quartiles are the three values that divide the data set into four equal groups. Again, the numbers are put in numerical order.

The median (Q_2, the **second quartile**) is the first to be calculated.

The median splits the data set into two equal groups.

- The **lower quartile** (Q_1) is the median of the lower half of the numerical data.

- The **upper quartile** (Q_3) is the median of the upper half of the numerical data.

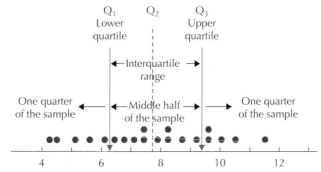

Example 7.3

a   These are the weights (in kg) of the boys in a youth club football team:

   60   60   62   64   65   65   66   69   70   71   83

   Calculate the median and quartiles of the weight of the team.

b   These are the weights (in kg) of the boys in the opposing football team (including their reserve):

   55   57   58   60   60   61   61   63   66   68   70   73

   Calculate the median and quartiles of the weight of the opposing team.

(continued)

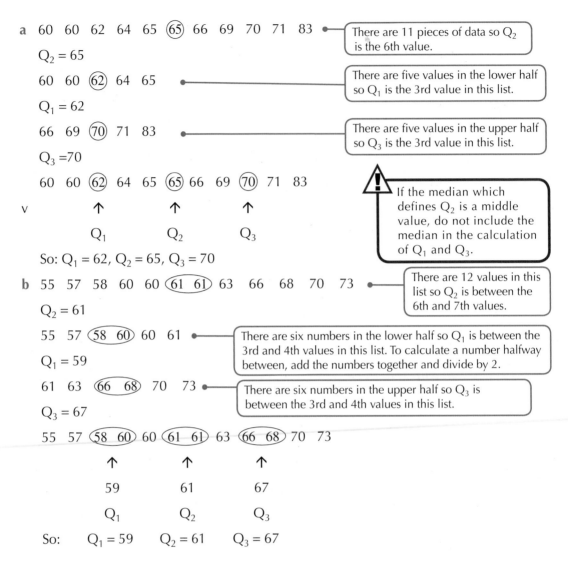

a 60 60 62 64 65 ⑥⑤ 66 69 70 71 83 •——— There are 11 pieces of data so Q_2 is the 6th value.

$Q_2 = 65$

60 60 ⑥② 64 65 •——— There are five values in the lower half so Q_1 is the 3rd value in this list.

$Q_1 = 62$

66 69 ⑦⓪ 71 83 •——— There are five values in the upper half so Q_3 is the 3rd value in this list.

$Q_3 = 70$

60 60 ⑥② 64 65 ⑥⑤ 66 69 ⑦⓪ 71 83

v   ↑   ↑   ↑

Q_1 Q_2 Q_3

⚠ If the median which defines Q_2 is a middle value, do not include the median in the calculation of Q_1 and Q_3.

So: $Q_1 = 62$, $Q_2 = 65$, $Q_3 = 70$

b 55 57 58 60 60 ⑥① ⑥① 63 66 68 70 73 •——— There are 12 values in this list so Q_2 is between the 6th and 7th values.

$Q_2 = 61$

55 57 ⑤⑧ ⑥⓪ 60 61 •——— There are six numbers in the lower half so Q_1 is between the 3rd and 4th values in this list. To calculate a number halfway between, add the numbers together and divide by 2.

$Q_1 = 59$

61 63 ⑥⑥ ⑥⑧ 70 73 •——— There are six numbers in the upper half so Q_3 is between the 3rd and 4th values in this list.

$Q_3 = 67$

55 57 ⑤⑧ ⑥⓪ 60 ⑥① ⑥① 63 ⑥⑥ ⑥⑧ 70 73

↑   ↑   ↑

59   61   67

Q_1 Q_2 Q_3

So: $Q_1 = 59$ $Q_2 = 61$ $Q_3 = 67$

Interquartile range

Using Q_1 and Q_3, the interquartile range (IQR) is calculated as follows:

$$IQR = Q_3 - Q_1$$

The IQR is a measure of spread giving the range of the middle 50% of the data.

- A small IQR means that the middle values are closer together and are more consistent.

- A larger IQR means that the middle values are further apart and are less consistent.

Making a comparison between two data sets using the median and interquartile range is similar to making a comparison using mean and standard deviation.

Example 7.4

a Calculate the interquartile range for:

   i the group of boys in the football team in Example 7.3 part **a**

   ii the group of boys in the opposing team in Example 7.3 part **b**.

b Make two valid comparisons between the weights of the two groups of boys.

a i   IQR = $Q_3 - Q_1$

          = 70 − 62

          = 8 kg

  ii   IQR = $Q_3 - Q_1$

          = 67 − 59

          = 8 kg

b   On average the football team weighed more than their opponents.

   The spread of weights of the boys in each team was equal.

Semi-interquartile range

Sometimes the **semi-interquartile range** (SIQR) is used instead of the interquartile range. It is calculated as follows:

$$\text{SIQR} = \frac{Q_3 - Q_1}{2}$$

The semi-interquartile range is half of the interquartile range, so it is a measure of the average spread from the centre of the data. This can be used as a reasonably good comparison with the standard deviation but tends to be used as a valid comparison if the data set is sufficiently large.

Any comparisons in SIQR between two data sets are answered in the same way as comparisons using interquartile range.

Example 7.5

A group of eight friends went ten-pin bowling one evening. At the end of the game their points were:

25   87   112   54   24   36   64   120

a   Calculate:

   i   the median

   ii   the quartiles

   iii   the semi-interquartile range of the points.

   The same group went ten-pin bowling once a week for 3 months. At the end of the game they played after the 3 months, their points were:

   65   112   101   90   143   105   98   121

b   Calculate:

   i   the median

   ii   the quartiles

   iii   the semi-interquartile range of the points.

c   Make two valid comparisons between the performance at the start and at the end of the 3 months.

a 24 $\widehat{25\ 36}$ $\widehat{54\ 64}$ $\widehat{87\ 112}$ 120 ←───[Rearrange the numbers in order (smallest to largest).]

   ↑   ↑   ↑

  30.5   59   99.5

   Q_1 Q_2 Q_3

  i $Q_2 = 59$

  ii $Q_1 = 30.5$, $Q_3 = 99.5$

  iii $\text{SIQR} = \dfrac{Q_3 - Q_1}{2}$

$$= \dfrac{99.5 - 30.5}{2}$$

$$= 34.5$$

b 65 $\widehat{90\ 98}$ $\widehat{101\ 105}$ $\widehat{112\ 121}$ 143

   ↑   ↑   ↑

  94   103   116.5

  Q_1 Q_2 Q_3

  i $Q_2 = 103$

  ii $Q_1 = 94$, $Q_3 = 116.5$

  iii $\text{SIQR} = \dfrac{Q_3 - Q_1}{2}$

$$= \dfrac{116.5 - 94}{2}$$

$$= 11.25$$

c   On average the results were higher after 3 months' practice.

  There was less variation in results at the end of the 3 months.

Exercise 7B

1   Find the median, quartiles and interquartile range of the following data sets:

  a   1, 1, 3, 3, 3, 6, 7, 7, 8, 9, 9

  b   23, 25, 29, 29, 32, 36, 41, 42, 42, 51, 52, 60

  c   4.5, 3.2, 1.6, 8.9, 5.3, 2.7, 8.4, 7.2, 5.6

2   A safari park has African elephants and Indian elephants.

  The park ranger measures the lengths of the ears of the African elephants. They have the following measurements in centimetres:

  115   145   163   135   128   131   165   124   181   172   118

  a   For the African elephants' ears calculate:

    i   the median

    ii   the quartiles

    iii   the interquartile range.

    b The park ranger also measures the lengths of the ears of the Indian elephants. They have the following measurements in centimetres:

31 43 56 38 60 41 38 34 52 48

    For the Indian elephants' ears calculate:

      i the median

      ii the quartiles

      iii the interquartile range.

    c Make two valid comparisons between the lengths of ears of African and Indian elephants at the safari park.

3 An estate agent has an office in Edinburgh and Glasgow.

The prices of some 2-bedroomed flats for sale in Edinburgh are:

£115 000 £173 000 £155 000 £128 000 £163 000 £141 000

£138 000 £167 000 £124 000 £153 000 £149 000

    a For these flats calculate:

      i the median

      ii the quartiles

      iii the interquartile range.

    b The prices of some 2-bedroomed flats for sale in Glasgow are:

£121 000 £138 000 £95 000 £108 000 £132 000 £99 000

£118 000 £136 000 £112 000 £85 000 £167 000

    For the flats in Glasgow calculate:

      i the median

      ii the quartiles

      iii the interquartile range.

    c Make two valid comparisons between the price of 2-bedroomed flats for sale in Edinburgh and Glasgow sold by this estate agent.

4 A rugby team participated in a rowing challenge and had to row as many strokes as possible in one minute. Their results were:

24 35 19 29 32 36 28 25 34 22 18 26 31 37 24

    a Calculate:

      i the median

      ii the quartiles

      iii the interquartile range.

    b Each player then followed a training programme for 8 weeks. At the end of the training programme they participated in the same challenge. Their results were:

26 38 18 32 35 38 30 25 39 41 31 27 35 36 29

    Calculate:

      i the median

      ii the quartiles

      iii the interquartile range.

    c Make two valid comparisons of the results before and after the 8-week training period.

★ 5 A hotel in Edinburgh has 25 rooms. The hotel manager records the number of rooms occupied per night during a 2-week period in August. These are her results:

18 25 16 17 20 15 18 19 23 25 19 17 18 20

a Calculate:

i the median

ii the quartiles

iii the interquartile range.

b The hotel manager also records the number of rooms occupied per night during a 2-week period in November. These are her results:

18 4 7 9 8 3 15 13 0 8 6 9 10 16

Calculate:

i the median

ii the quartiles

iii the interquartile range.

c Make two valid comparisons between the numbers of rooms occupied per night in August and November.

d If you were the manager of this Edinburgh hotel, what decisions would you make based on this information?

6 A popular newspaper ran a series of articles on 'The world's fattest country' using data provided by the United Nations.

They measured and recorded these weights (in kg) of seven male US citizens:

86.3 78.1 91.7 74.5 95.6 69.7 84.4

a Calculate:

i the median

ii the quartiles

iii the interquartile range.

b They also measured and recorded these weights (in kg) of seven male Mexican citizens:

87.2 79.1 70.6 85.2 92.6 75.1 96.8

b Calculate:

i the median

ii the quartiles

iii the interquartile range.

c Make two valid comparisons between the weights of the US and Mexican citizens in this sample.

7 Jenny is 20 years old and has a five-year-old Fiat 500. She needs to insure her car and researches the best car insurance available. She contacts six different insurance companies who give her these annual quotes:

£2200 £2950 £2450 £2880 £2286 £2812

a Calculate:

i the median

ii the quartiles

iii the interquartile range.

b Jenny's sister Georgia is 27 years old and has the same type of car. Georgia also obtains some car insurance quotes. The median quote she receives is £1850 and the interquartile range is £900.

Write two valid comparisons between the car insurance costs for Jenny (a 20 year old) and Georgia (a 27 year old).

8 A GP records the ages (in years) of the patients who visit him one day. The stem and leaf diagram shows his results.

0	1 2
1	5
2	7
3	3 3 6
4	2
5	3 8
6	0 3 5 6
7	2 5 8
8	2 6 7 9

$n = 21$ $2 \mid 6 = 26$

a Find the median and quartiles.

b Calculate:

  i the range

  ii the interquartile range.

c The GP was later called out to another patient aged 101.

  i Calculate the new range and interquartile range

  ii Which is more affected by the addition of this piece of data: the range or the interquartile range?

> ⚠️ The interquartile range only calculates the range of the middle 50% of the data, so including an outlying point in the data does not affect the interquartile range.

9 A popular new concert venue has an audience capacity of 12 000 people.

The audience figures for 12 music concerts were:

8125   10 123   9138   11 325   10 036   9108

7138   11 950   9367   10 318   11 143   8749

a Calculate:

  i the median

  ii the quartiles

  iii the semi-interquartile range.

b The venue is also used for major sports events. The audience figures for 12 sports events were:

9991   9109   11 546   10 435   11 645   10 739

11 876   10 259   10 297   11 421   9709   11 988

Calculate:

  i the median

  ii the quartiles

  iii the semi-interquartile range.

c Based on these figures which do you think is more popular: music concerts or sports events? Give two reasons for your answer.

10 Janie and her family were on holiday in France and Spain. Janie recorded the cost of a glass of coke (in euros, €) in France. Her results were:

1.65 2.50 2.80 1.90 2.10 2.90 1.75 1.95 2.25

a Calculate:

   i the median

   ii the quartiles

   iii the semi-interquartile range.

b Janie also recorded the cost of a glass of coke (in €) in Spain. Her results were:

1.60 1.80 1.95 1.75 2.00 2.10 1.95 2.20 1.70 1.85

Calculate:

   i the median

   ii the quartiles

   iii the semi-interquartile range.

c Make two valid comparisons between the cost of a glass of coke in France and in Spain.

Boxplots

You can draw a **boxplot** if you want to show the median and quartiles in a diagram.

In order to draw a boxplot you need to state a **five-figure summary**. This consists of:

- lowest number, **L**
- lower quartile, Q_1
- median, Q_2
- upper quartile, Q_3
- highest number, **H**

To draw the diagram:

- draw a graduated number line with a label clearly shown under the number line
- draw five vertical lines above the line to represent each element of the five-figure summary
- draw a box round the upper and lower quartiles
- join the box to the lowest and highest numbers with a horizontal line.

The diagram below illustrates the features of a boxplot.

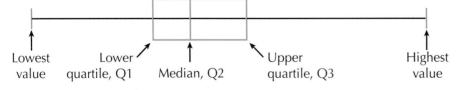

Lowest value   Lower quartile, Q1   Median, Q2   Upper quartile, Q3   Highest value

The boxplot below is drawn for the data in Example 7.3 part **a**.

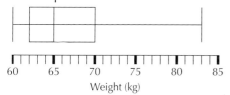

60   65   70   75   80   85
Weight (kg)

Range = Highest value − lowest value

You may also be given information on a boxplot and asked to make a comparison using the data from the boxplot. The median and interquartile range can be extracted and compared, and the boxplot itself can be used as a comparison of the range.

Example 7.6

The results in last year's S3 maths exam are shown in the boxplots below.

The results for the boys and girls have been shown in separate boxplots.

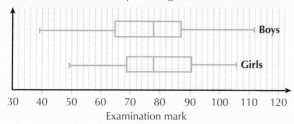

30   40   50   60   70   80   90   100   110   120
Examination mark

a   Write down the median of:

   i   the boys' results

   ii   the girls' results.

b   Calculate the interquartile range of:

   i   the boys' results

   ii   the girls' results.

c   Comment on the boys' and girls' results.

a   i   Median = 78   ii   Median = 78

b   i   IQR = 87 − 65 = 22   ii   IQR = 91 − 69 = 22

c   On average the boys' and girls' results were similar and the spread of the middle 50% was the same, but the boys results were far more spread out than the girls as the range was much bigger.

Exercise 7C

★   1   For each set of data:

   i   create a five-figure summary and draw a boxplot

   ii   calculate the interquartile range.

      a   3, 8, 8, 9, 12, 14, 15, 18, 21

      b   15, 16, 18, 19, 19, 22, 23, 23, 24, 26, 27, 29, 30, 33

      c   25, 22, 21, 27,35, 38, 33, 25, 29, 27, 28, 34, 24, 28, 33, 27, 25,29,31

      d   101, 110, 105, 123, 113, 105, 109, 118, 120, 114

2   The boxplot shows the times taken for a group of pensioners to do a set of long-division calculations.

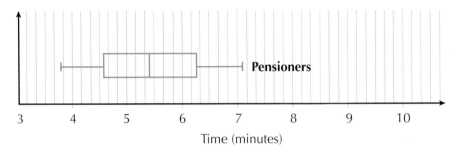

Time (minutes)

The same set of calculations were given to some S3 students. Their results were:

shortest time: 3 minutes 20 seconds

lower quartile: 6 minutes 10 seconds

median: 7 minutes

upper quartile: 7 minutes 50 seconds

longest time: 9 minutes 40 seconds.

a   Copy the diagram and draw a boxplot to show the students' times.

b   Comment on the differences between the two distributions.

3   The table shows the five-figure summary for the annual salaries of 200 men and women.

	Lowest	Lower quartile	Median	Upper quartile	Highest
Women	£8000	£15 000	£17 000	£19 500	£31 000
Men	£7000	£18 000	£21 000	£25 000	£43 500

a   Draw a boxplot for men's salaries and a boxplot for women's salaries on the same diagram.

b   Make two valid comparisons between the salaries of men and women, linking your comments to the information in the boxplots.

4   Three different fertilisers were used in a study of tomato plants. The table shows the heights of the plants (in cm) after 5 weeks.

	Lowest (cm)	Lower quartile (cm)	Median (cm)	Upper quartile (cm)	Highest (cm)
Fertiliser A	20	26	30	32	35
Fertiliser B	21	24	25	26	27
Fertiliser C	24	26	32	33	36

a   Draw boxplots to illustrate these results on the same diagram.

b   Comment on the effectiveness of the three fertilisers.

5   The table shows the average maximum monthly temperatures (in °C) in the UK and Spain.

	Jan (°C)	Feb (°C)	Mar (°C)	Apr (°C)	May (°C)	June (°C)	July (°C)	Aug (°C)	Sept (°C)	Oct (°C)	Nov (°C)	Dec (°C)
UK	6	9	10	13	16	19	21	23	19	15	8	2
Spain	9	12	14	16	21	26	31	30	26	19	13	10

a   Draw boxplots for the temperatures in Spain and the UK on the same diagram.

b   Make two valid comparisons between the temperatures, linking your comments to the information in the boxplots.

★ 6   A survey recorded the length of time cars were parked in a town centre. The survey took place on a Friday and a Saturday. The boxplot shows the results of the survey.

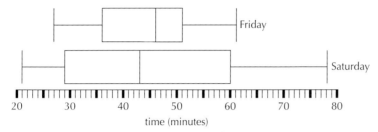

a   Use the boxplots to make a five-figure summary of the times recorded for each day.

b   Calculate the semi-interquartile range for each day.

c   Make two valid comparisons between the lengths of times people parked on the Friday and the Saturday.

7   A GP practice has two doctors, Dr Ball and Dr Charlton. The boxplots illustrate the waiting times (in minutes) for each doctor's patients during November.

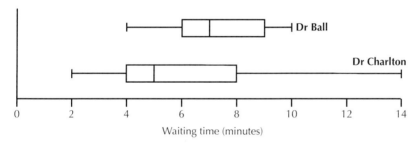

Hannah was deciding which doctor to see. Which one would you advise her to see and why? Link your comments to the information on the boxplots.

8   The boxplots show information about the midday temperatures in two holiday towns in September.

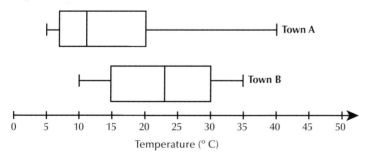

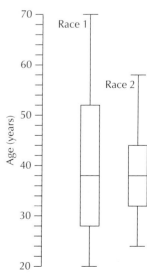

Compare the temperatures in towns A and B in September. Link your comments to the information on the boxplots.

9   A popular race takes place each year. Participants can run in the 10 km race or the half-marathon (21 km). The boxplots show the ages of the participants of the two races.

a   Use the boxplots to make a five-figure summary of the ages of participants.

b Calculate the interquartile range of the age for each race.

c Which race do you think is the 10 km race and which is the half-marathon? Give a reason for your answer.

d Make two valid comparisons between the ages of participants in the 10 km and the half-marathon.

10 A ski resort in the USA and a ski resort in Italy recorded their daily snowfall (in centimetres) during January and February. The boxplots show their results.

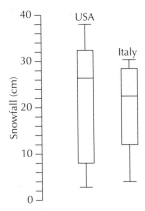

a Calculate the median and the semi-interquartile range of snowfall for each ski resort.

b Which ski resort would you visit for a more consistent fall of snow in January or February? Give a reason for your answer.

11 Four female runners are selected to represent their country in the next athletics championships in the 400 m relay race. The following boxplots show their race times (in seconds) in 400 m races over the past 12 months.

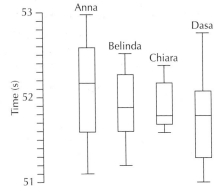

The coach wants to put the fastest runner on the last leg of the relay and the second fastest on the first leg. Which athletes should she pick for the first and last legs? Give reasons for your answers.

12 Rodrigo was given a diagram showing boxplots for the amount of daily sunshine in the resorts of Bude and Torquay for August, but no scale was given on the diagram. He was told to write a report on the differences between the amounts of sunshine in the two resorts.

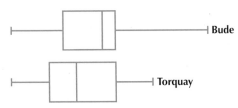

Create a report that he could reasonably write using these boxplots even though no scale was given.

Which measure of spread should I use?

When you are answering questions in an exam it is obvious which measure of spread to use because the question will direct you. However, in other circumstances you may not be given any clues, so it is important that you understand and recognise which measure would be most appropriate.

Use **standard deviation** if there are no outlying points, the data set is relatively small and you want a more accurate measure of the spread.

Use **interquartile range** if there are outlying points and the data set is relatively large.

Exercise 7D

★ 1 The table shows the total number of recorded crimes in six countries for 2011.

Country	Germany	Spain	Italy	Netherlands	France	Canada
Number of reported crimes	2112843	377965	900870	372305	1172547	628920

Source: UN Office on Drugs and Crime

a i Find the mean and standard deviation of the total number of crimes reported in these six countries correct to 1 decimal place.

ii Find the median and interquartile range of the total number of crimes reported in these six countries.

The total number of crimes reported in the USA was 12408899.

b i Calculate the mean and standard deviation of the total number of crimes reported in all seven countries, correct to 1 decimal place.

ii Calculate the median and interquartile range of the total number of crimes reported in all seven countries.

c i What impact has the addition of the USA's statistic to the group of countries had on the mean and standard deviation?

ii What impact has the addition of the USA's statistic to the group of countries had on the median and interquartile range?

iii Which measure is affected more by the inclusion of the USA's statistic?

2 The annual salaries for a company's employees are:

£83000   £65000   £34000   £28000   £28000
£20000   £20000   £20000   £8000

a Calculate:

i the mean salary

ii the median salary

iii the modal salary (the mode)

iv the standard devation

v the interquartile range.

> The mode is the most popular value. ⚠

b The company's general manager suggests a pay rise of 6% per employee. Calculate:

   i the new mean salary

   ii the new median salary

   iii the new modal salary (the mode)

   iv the new standard devation

   v the new interquartile range.

c The workers suggest a pay rise of £1500. Calculate:

   i the new mean salary

   ii the new median salary

   iii the new modal salary (the mode)

   iv the new standard devation

   v the new interquartile range.

d Which pay rise has a bigger effect on:

   i the standard devation

   ii the interquartile range?

e If you were the general manager, which would you use to show that the salaries are better and more consistent: mean and standard deviation or median and interquartile range?

f Why do you think that the general manager might be opposed to the pay rise of £1500? Give a reason for your answer.

3 Two families take part in a tug o'war competition. The table shows the members of each family and their weights.

Key family		Charlton family	
Name	Weight (kg)	Name	Weight (kg)
Brian	58	David	60
Ann	32	Hannah	56
Steve	49	Pete	42
Alison	39	Barbara	76
Jill	64	Chris	71
Holly	75	Julie	39
Albert	52	George	22

a Each family have to choose four members with a mean weight of 45–50 kg. Choose two teams, one from each family.

b Calculate the standard deviation of the two teams you have selected and make a valid comparison between the two teams.

c Now choose two teams, one from each family, if they have to choose four members with a median weight of 45–50 kg.

d Have you chosen the same or different teams for each family? Give reasons for your answer.

e Now calculate the interquartile range of each of the teams chosen in part c.

f Which team would each family prefer: the one with a mean of 45–50 kg or the one with the median of 45–50 kg? Give reasons for your answers.

4   The table shows the marks (in %) of 10 students in Papers 1 and 2 of their National 5 Mathematics Prelim Exam.

Student	Ann	Bridget	Carole	Daniel	Ed	Fay	George	Heidi	Imman	Jez
Paper 1	72	61	43	92	56	62	73	56	38	67
Paper 2	81	57	49	85	62	61	70	66	48	51

a   i   Calculate the mean and standard deviation of each paper.

   ii   Using this information make two valid comparisons between the two papers.

   iii How many students scored above the mean in both papers?

b   i   Calculate the median and interquartile range of each paper.

   ii   Using this information make two valid comparisons between the two papers.

   iii How many pupils scored above the median in both papers?

c   Now compare your answers to **a** and **b**. Which set of results shows that the papers are similar in difficulty? Give a reason for your answer.

d   If you were the class teacher, which measure of average and spread would you quote to the class for:

   i   Paper 1

   ii   Paper 2?

   Give reasons for your answers.

5   The following table shows the marks (out of 10) for five couples in a dancing competition.

Dance	Couple				
	Kath and Brian	Tom and Helen	Joe and Nic	Azan and Phyllis	David and Heather
Tango	10	6	4	8	6
Salsa	6	8	3	8	6
Ballroom	8	4	4	8	8

a   Kath said she thought the Salsa was the hardest dance. Calculate the mean score for each dance and use these to decide whether or not Kath was correct.

b   Now calculate the median score for each dance. Use this information to decide if Kath was correct.

c   Calculate the standard deviation for each dance and use this to state which dance provided the most consistent marks.

d   Why would it not be appropriate to use the interquartile range for these data sets?

> The interquartile range needs a reasonable amount of data for a meaningful comparison to be made.

GO! Activity

10 boys in an S4 class were asked to measure their height, hand span, shoe size and arm span. The table shows their results.

Boy	Height (cm)	Hand span (cm)	Shoe size	Arm span (cm)
Adam	170	20.2	9	160
Calum	165	18.5	9	154
Jack	176	21.5	11	164
Jonny	154	17.5	8	147
Steven	168	19.5	10	156
Charlie	172	20.3	11	163
Sean	175	22.1	12	167
Archie	164	18.7	9	151
Gregor	160	18.4	8	149
Harry	176	21.7	10	168

1   Calculate the mean, median, standard deviation and interquartile range of the data in the table.

2   Measure the height, hand span, shoe size and arm span of 10 15-year-old boys at your school and compare their results to those in the table.

3   Now measure the height, hand span, shoe size and arm span of 10 15-year-old girls at your school and compare their results to the boys' results you collected.

4   Draw boxplots to compare the results of the 15-year-old boys and girls at your school for the different categories and use your boxplots to make comparisons.

- I can analyse and compare two or more data sets using mean and standard deviation. ★ Exercise 7A Q5 ★ Exercise 7D Q1

- I can analyse and compare two or more data sets using median and interquartile range. ★ Exercise 7B Q5 ★ Exercise 7D Q1

- I can construct and interpret a boxplot. ★ Exercise 7C Q1, Q6

For further assessment opportunities, see the Case Studies for Unit 1 on pages 136–140.

8 Drawing a best-fitting line for given data

This chapter will show you how to:

- draw a best-fitting line for a set of data
- use the best-fitting line to make predictions
- analyse and compare two or more data sets using scatter graphs.

You should already know:

- how to construct a scatter graph
- how to recognise the trend and correlation of a scatter graph.

Scatter graphs

Scatter graphs are used to investigate the relationship between **two variables.** Corresponding values are plotted on a graph in the same way as coordinates are plotted.

When changes in one variable match changes in the other, there may be a cause-and-effect relationship between the two. This is called the **correlation.** Correlation can be described in three ways.

- There is a **positive correlation** when the results show that, as one quantity increases, the other quantity also increases.

- There is a **negative correlation** when the results show that, as one quantity increases, the other quantity decreases.

- There is **no correlation** when there is no relation between the two data sets.

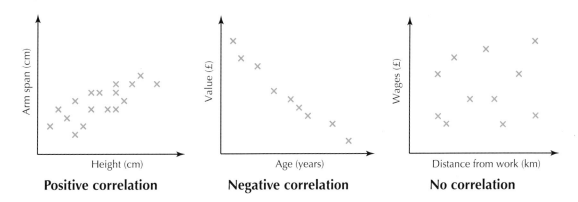

| **Positive correlation** | **Negative correlation** | **No correlation** |

For example, the conclusion to be drawn from the positive correlation diagram is, 'The taller people are, the wider their arm span is likely to be.'

The conclusion from the negative correlation diagram is, 'The older a car is, the lower its value is likely to be.'

The conclusion from the no correlation diagram is, 'The distance you live from your place of work is unlikely to affect how much you earn.'

Best-fitting line

A **best-fitting line** can be drawn on a scatter graph which has either a positive or a negative correlation. There are a few basic guidelines for drawing a best-fitting line:

- the line should be as close as possible to all the points
- ensure you have approximately the same number of points on either side of the line and that the line is following the direction of the data
- the best-fitting line might not go through the first and last data points on the graph.

The line can then be used to predict a related value.

Example 8.1

Pupils at a school demonstrated how many press-ups and sit-ups they could do in 3 minutes. The scatter graph represents the information.

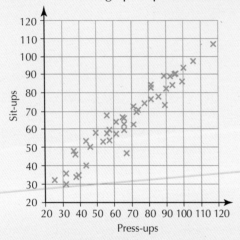

a   Describe the correlation of the graph and what the graph tells you.

b   Draw a best-fitting line on the graph.

c   Qasim was absent on the day of the sit-ups exercise but was in class when they did the press-ups exercise. He did 75 press-ups. Use your best-fitting line to predict how many sit-ups he would be able to do in 3 minutes.

a   The graph shows a positive correlation: the more press-ups a person can do, the more sit-ups they can also do.

b

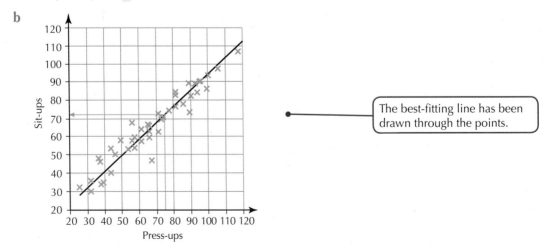

The best-fitting line has been drawn through the points.

c

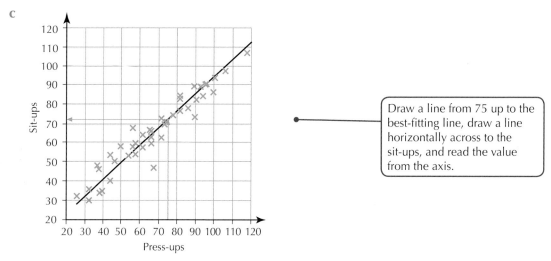

Draw a line from 75 up to the best-fitting line, draw a line horizontally across to the sit-ups, and read the value from the axis.

It is predicted that Qasim could do 72 sit-ups.

Exercise 8A

1 Describe the correlation of each of these graphs, and state what each graph tells you.

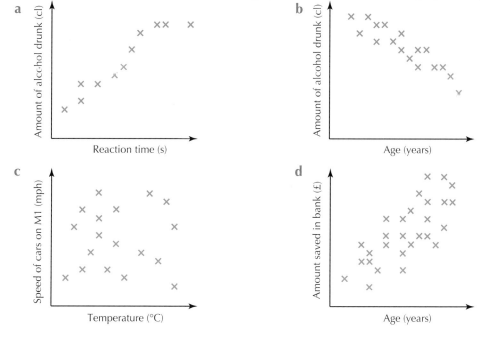

2 The scatter graph shows the relationship between the total mileage of a car and its value as a percentage of its original value.

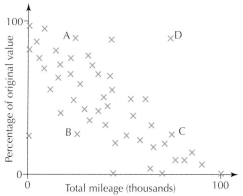

a **i** Which of the four points A, B, C or D represents each of these statements.

Rhona: 'I have a rare car. It has done a lot of miles but it is still worth a lot of money.'

Haider: 'My car is quite new. It hasn't done many miles.'

Jim: 'My car hasn't done many miles, but it is really old and rusty.'

ii Write a statement by Jennifer to match the fourth point.

b What does the graph tell you about the relationship between the mileage of a car and the percentage of its original value?

3 Sketch a scatter graph which shows the relationship between the amount of petrol used and the distance driven.

4 Ten pupils sat prelim exams in Maths and Physics. The table below shows their marks out of 100.

| Maths marks | 51 | 63 | 73 | 20 | 45 | 90 | 32 | 58 | 55 | 68 |
| Physics marks | 32 | 50 | 61 | 10 | 40 | 76 | 28 | 41 | 48 | 52 |

a Draw a scatter graph to illustrate the data in the table.

b Describe the correlation and the connection between the Maths and the Physics marks.

c Draw a best-fitting line on your diagram.

d Use your line to estimate the Physics mark of a pupil who achieved a Maths mark of 40.

e Use your line to estimate the Maths mark of a pupil who achieved a Physics mark of 65.

f Which exam do you think was easier? Give a reason for your answer.

5 The table below shows the scores given by two judges in an ice-skating competition.

| Judge 1 | 5.8 | 5.2 | 4.5 | 4.7 | 5.6 | 5.9 | 5.0 | 5.8 | 4.3 |
| Judge 2 | 5.6 | 4.9 | 4.4 | 4.3 | 5.2 | 5.7 | 5.1 | 5.7 | 4.3 |

a Draw a scatter graph to illustrate the data in the table.

b Describe the correlation and what the graph tells you.

c Draw the best-fitting line.

d Judge 1 gives the next ice-skater a mark of 4.9. Use the best-fitting line to estimate the mark that Judge 2 is likely to give.

e Judge 2 gives another ice-skater a mark of 5.4. Use the best-fitting line to estimate the mark that Judge 1 is likely to give.

★ **6** A general practitioners' clinic is doing a study into the ages of babies and number of hours they sleep. The table shows the ages of nine babies under one year old and the average number of hours they sleep on a particular day.

| Age (years) | 0.25 | 0.18 | 0.67 | 0.85 | 0.34 | 0.55 | 0.72 | 0.91 | 0.88 |
| Amount of sleep (hours) | 15.3 | 15.1 | 13.4 | 13.1 | 15.1 | 14.7 | 14.1 | 13.1 | 13.6 |

a   Draw a scatter graph to illustrate the data in the table

b   Describe the correlation and what the graph tells you.

c   Draw the best-fitting line.

d   Use the best-fitting line to estimate the length of time a child aged exactly 6 months is likely to sleep.

> ⚠️ Age is shown as a decimal fraction in the table. This means that 6 months is represented as 0.5 years.

7   Nine children took part in a study into the connection between their reading speed in words per minute (wpm) and their intelligence quotient (IQ). Their results are shown in the table.

Name	Ali	Brian	Charles	David	Emma	Hannah	Kate	Joanne	Frank
Reading speed (wpm)	130	110	180	170	200	230	140	190	160
IQ	95	90	110	105	115	120	100	110	105

a   Draw a scatter graph to illustrate the data in the table.

b   Describe the correlation and what the graph tells you.

c   Draw the best-fitting line.

> ⚠️ IQ is one measure of intelligence.

d   Use the best-fitting line to estimate the reading speed of a child with an IQ of 85.

e   Use the best-fitting line to estimate the IQ of a child with a reading speed of 210 wpm.

8   The cost of a 7-day holiday and the length of time (in hours) taken to fly to the destination are compared for 10 holidays. The results are shown in the table.

Flight time (hours)	2	2.5	2.5	2.5	3	3	4	4.5	6	8
Cost of holiday (£)	188	142	152	158	170	237	259	242	358	510

a   Draw a scatter graph to illustrate the data in the table.

b   Draw a best-fitting line on the diagram.

c   Use your line to estimate the cost of a holiday if the flight lasts 5 hours.

d   Use your line to estimate the length of time the flight is expected to take if the holiday costs £310.

9   During a study week prior to examinations, pupils were asked how many hours they spent revising and how many hours they spent on the computer for leisure (watching films, playing games, emailing or on social networking sites). The table shows the results.

Revision time (hours)	15	20	22	28	35	38	40	43	50
Computer time (hours)	38	28	30	25	15	15	14	10	8

a   Draw a scatter graph to illustrate the data in the table.

b   Draw a best-fitting line on the diagram.

c   Use your line to estimate the length of time spent on the computer if a pupil studied for 32 hours.

d   Use your line to estimate the number of hours of revision if a pupil spent 20 hours on the computer that week.

4   Frank was on holiday in Rome. He noticed that a 500 ml bottle of water seemed to be expensive near the Vatican City but was less expensive elsewhere. He decided to compare the prices of 500 ml bottles of water at different distances from the Vatican City. He recorded the prices and the table below shows his results.

Distance from Vatican City (km)	1	2	3	4	5	6	7	8	9	10
Price of bottle of water (€)	2.10	1.80	1.90	1.50	1.40	1.20	1.00	0.80	0.70	0.70

a   Draw a scatter graph to illustrate the data and draw a best-fitting line.

b   How could this type of study help tourists in Rome and what advice could Frank give them?

c   Frank's friend Gina was also in Rome on holiday. In preparation for her holiday she researched the cost of a room per night in a two-star hotel in Rome and its distance from the Vatican City. The table shows her results.

Distance from Vatican City (km)	1	2	3	4	5	6	7	8	9	10
Cost of hotel room per night (€)	100	94	89	87	82	79	77	72	68	61

Draw a scatter graph to illustrate the data and draw a best-fitting line.

d   Compare the scatter graphs you have drawn. What conclusion can you reach about the relationship between the cost of water and accommodation and their distance from the Vatican City? Give reasons for your answer.

★ 5   The table shows the age and height of 10 Jamaican 100 m sprint athletes.

Age (years)	11.9	12.4	12.8	13.2	13.5	13.7	14.4	14.7	14.7	15.2
Height (m)	1.52	1.54	1.62	1.61	1.68	1.72	1.77	1.79	1.88	1.85

a   Plot the data on a scatter graph.

b   Draw a best-fitting line on the graph.

c   Use the best-fitting line to estimate the height of a Jamaican 100 m sprinter who is aged 14.1 years.

d   Use the best-fitting line to estimate the age of a Jamaican sprinter whose height is 2.2 m. Suggest why this might not be correct.

e   The table shows the age and height of 10 Jamaican 100 m sprinters aged over 18.

Age (years)	24.3	26.9	28.6	25.3	22.8	25.7	22.1	24.6	27.7	29.1
Height (m)	1.92	2.05	1.98	2.16	2.03	1.96	2.21	1.99	2.09	1.82

Plot the data on a scatter graph and draw a best-fitting line on the graph.

f   Compare the scatter graphs you have drawn in parts **a** and **e**. What conclusions might you come to about the age and height of 100 m sprinters?

★ 6 Gross domestic product (GDP) is a measure of a country's economic growth. GDP per capita is the GDP divided by the population. The table shows the GDP per capita and average life expectancy of some countries in Western Europe.

Country	GDP per capita (€000s)	Life expectancy (years)
Austria	29.1	77.48
Belgium	32.5	78.29
Finland	28.4	77.32
France	33.3	78.63
Germany	28.4	77.17
Ireland	27.3	76.39
Italy	33.1	78.51
Netherlands	31.7	78.15
Switzerland	34.2	78.99
United Kingdom	28.9	77.37

a Plot the data on a scatter graph and draw a best-fitting line on the graph.

b Use your graph to estimate the life expectancy of a Western European country whose GDP per capita (€000s) is recorded as 29.5.

c The table shows the GDP per capita (given in €s to make it easier to compare) and average life expectancy for some countries in Africa.

Country	GDP per capita (€000s)	Life expectancy (years)
Burkina Faso	0.92	44.46
Burundi	0.5	43.20
Congo (Republic of)	0.6	50.02
Ethiopia	0.57	41.24
Guinea-Bissau	0.68	46.97
Mali	0.75	45.43
Kenya	0.8	45.22
Malawi	0.52	37.98
Niger	0.8	51.01
Sierra Leone	0.42	42.84
Somalia	0.42	77.37
Tanzania	0.5	44.56
Zambia	0.7	35.25

Plot the data on a scatter graph and draw a best-fitting line on the graph.

d Compare the scatter graphs you have drawn in parts **a** and **c**. What conclusions might you come to about the relationship between GDP and life expectancy in Western Europe and Africa?

7 A class of 30 pupils sat an exam in January. Between the exam in January and the final exam in May, their teacher ran a revision class to help them with the final exam. 15 of the pupils attended the revision class.

The table shows the results of the pupils who did not attend the revision class.

January exam (%)	29	41	70	50	58	35	20	58	43	21	52	51	56	30	55
May exam (%)	27	40	72	49	59	39	18	61	47	24	52	54	55	34	58

a Plot the data on a scatter graph and draw a best-fitting line on the graph.

b The table shows the results of the pupils who attended the revision class.

January exam (%)	14	50	28	31	39	40	54	48	52	60	48	49	70	60	36
May exam (%)	18	58	40	40	50	56	70	63	67	75	65	62	86	76	48

Plot the data on a scatter graph and draw a best-fitting line on the graph.

c Compare the scatter graphs you have drawn in parts **a** and **b**. What conclusions might you come to about the effectiveness of the revision class? Give a reason for your answer.

8 A group of unfit men join a new 'boot camp' fitness club which meets twice a week. At the end of the first day's fitness club meeting, each man takes his pulse. The table shows the age of each man and his pulse rate (in beats per minute, bpm).

Age (years)	22	32	38	41	43	47	52	65
Pulse (bpm)	190	182	170	172	166	158	154	144

a Plot the data on a scatter graph and draw a best-fitting line on the graph.

b The men attend the club regularly for six months. At the end of a club meeting after six months, each man takes his pulse and records it again. The table shows the age of each man and his pulse.

Age	22	33	39	41	44	47	53	65
Pulse (bpm)	128	126	120	121	118	117	112	108

Plot the data on a scatter graph and draw a best-fitting line on the graph.

c Compare the scatter graphs you have drawn in parts **a** and **b**. What conclusions can you draw about the effectiveness of the fitness club? Give a reason for your answer.

GO! Activity

All pupils in the class should measure their handspan, shoe size, height and the distance they live from school and write down their results

As a class you can see whether there is a correlation between the sets of data:

- Choose two sets of data (e.g. handspan and shoe size) to compare.
- When the weather is suitable, as a class go outside and decide on your axes.
- Pupils should then stand in the correct place on the ground according to their two pieces of data.
- If it is possible, a teacher or other adult could photograph the results (perhaps from a window at least one floor from ground level) or record them to show back in the classroom at a later stage.

a Using this method, what is the correlation in your class between:

    i handspan and shoe size

    ii handspan and height

    iii handspan and distance from the school

    iv shoe size and height

    v shoe size and distance from the school

    vi height and distance from the school?

b If you have both boys and girls in your class, compare the correlation of the above data between boys' and girls' scatter graphs.

GO! Activity

1 a Research GDP and life expectancy of other continents.

  b Draw scatter graphs of these continents and comment on any correlation:

    i South America

    ii Asia.

  c i From your research, what are the causes of lower GDP and low life expectancy?

    ii What are the effects of lower GDP and low life expectancy?

2 a Research the life expectancy and associated fertility (number of children per woman) for different countries.

  b i What can you say about countries with the highest fertility? Give a reason for your answer.

    ii Are any countries an exception to this? Why do you think this is the case?

3 Investigate your own sets of data, for example:

- income vs children dying before the age of 5
- health spending vs life expectancy
- improved water access vs infant mortality rate.

- I can draw a best-fitting line for a set of data.
  ★ Exercise 8A Q6

- I can use the best-fitting line to make a prediction.
  ★ Exercise 8A Q10, ★ Exercise 8B Q2, Q5

- I can analyse and compare two or more data sets using scatter graphs. ★ Exercise 8B Q1, Q3, Q4, Q5

For further assessment opportunities, see the Case Studies for Unit 1 on pages 136–140.

9 Using a combination of statistical information presented in different diagrams

This chapter will show you how to:

- represent data graphically
- read and interpret data presented graphically.

You should already know:

- how to draw a bar graph
- how to draw a line graph
- how to draw a stem and leaf diagram
- how to draw a best-fitting line.

Representing data graphically

Statistics are used every day in business, on the news and in leisure activities. Statistics are used to help launch a new product, to predict who is going to win the next general election or to help decide on the best holiday destination based on what weather to expect in the resort.

Statistics can be presented in numerical form as averages and spread, or in graphical form. People often find graphical form easier to understand as they can see a 'picture' of what is happening.

This chapter deals with various graphical forms that can be used to represent data. These include bar graphs, histograms, line graphs, pie charts and scatter diagrams.

Example 9.1

The table shows the number of votes for each party in a local government election.

Party	Number of votes
Conservative	68
Liberal	23
Labour	235
SNP	308
Independent	103
Other	31
Total	768

Show this information in a pie chart.

(continued)

$$\frac{68}{768} \times 360° = 32° \text{ (nearest degree)}$$

Calculate the angle at the centre of the pie chart for each party. Use the number of votes for the Conservatives as a fraction of the overall votes and multiply by 360° to obtain the angle.

Party	Number of votes	Angle
Conservative	68	32°
Liberal	23	11°
Labour	235	110°
SNP	308	144°
Independent	103	48°
Other	31	15°
Total	**768**	**360°**

Check that the angles add to 360°. Remember that rounding each angle to the nearest degree will mean that the total might not add exactly to 360°.

Local election results

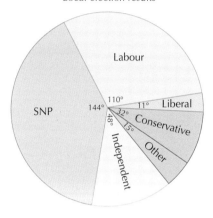

Example 9.2

The temperature and rainfall in Tenerife from April to September were recorded.

The table shows the results.

	April	May	June	July	August	September
Temperature (°C)	17	18	20	21	23	22
Rainfall (mm)	31	11	5	1	3	13

Illustrate this data in an appropriate diagram.

It is possible to combine two different types of diagram to show this information clearly. The temperature is probably best shown as a line graph and the rainfall is best shown as a bar graph.

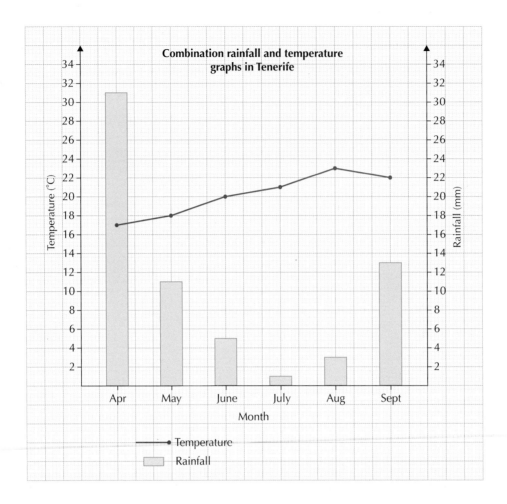

Combination rainfall and temperature graphs in Tenerife

Exercise 9A

1 Eight students each sat two assessments in maths, one in August and the other in the following May.

		Student							
		A	**B**	**C**	**D**	**E**	**F**	**G**	**H**
Result	**August**	30%	42%	35%	62%	71%	45%	48%	52%
	May	35%	59%	42%	70%	85%	38%	50%	52%

    a Draw a composite bar graph to illustrate these results.

    b State a valid comparison between the results for August and May.

★ **2** The table shows the IQ scores of 9 pupils and their reading speed in words per minute (wpm).

IQ score	105	110	100	130	80	95	95	115	120
Reading speed (wpm)	180	220	160	275	120	150	140	240	260

   a   Using appropriate scales, draw a scatter diagram to represent this data.

   b   Draw in the best-fitting line.

   c   Use your best-fitting line to predict the reading speed for a pupil with an IQ score of 90.

   d   Joanne predicts that a reading speed of 250 word/min will give an IQ score of 125. Is Joanne correct? Justify your answer.

3   Rose Street Primary School surveyed all the pupils in Primary 7 to find out how they travelled to school each day.

    There are 42 pupils in P7.

    8 boys said that they walked, 5 cycled, 3 came by bus and 7 came by car.

    10 girls said that they came by bus, 4 came by car, 3 walked and 2 cycled.

    Construct a table and a graph to illustrate these results.

4   The heights of the boys and girls in class 4Y2 were measured.

    Their heights, in cm, are shown.

    Girls: 154, 138, 152, 133, 141, 146, 157, 162, 102, 143, 126, 133, 151, 144, 137

    Boys: 163, 174, 155, 166, 172, 187, 161, 160, 186, 173, 187, 191, 176

    Illustrate this data on a back-to-back stem and leaf diagram.

    Compare the heights of the girls and the boys.

5   In one year a car showroom sells 23 red cars, 41 blue cars, 18 white cars and 35 black cars.

    Draw a pie chart to illustrate this information.

Reading and interpreting data presented graphically

It is often necessary to interpret the information given in a statistical diagram.

To do this you will need to understand the scale used on each axis as well as interpret the context of the graph.

Example 9.3

The owner of Beryl's Beauty Salon wants to analyse which of their four most popular treatments brings in the most money to the salon.

She prepares the following diagrams to show the data that she has collected over one week in the salon. 200 clients had the four most popular treatments.

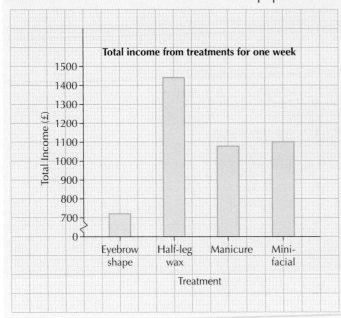

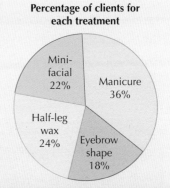

Percentage of clients for each treatment

Treatment	Time taken (min)
Eyebrow shape	10
Half-leg wax	25
Manicure	20
Mini-facial	30

a Calculate which treatment raises the most money per hour for the salon.

b What advice could you give to the manager so as to raise income for the salon?

a The table below uses the data given to show:

- the total amount of money taken by the salon for each treatment

- the number of clients who took each treatment

- the time each treatment takes.

Work out the income per treatment.

Work out the income per hour for each treatment.

Treatment	Total income (£)	Number of clients	Income per treatment (£)	Time per treatment (min)	Income per hour (£)
Eyebrow shape	720	36	720 ÷ 36 = 20	10	20 ÷ 10 × 60 = 120
Half-leg wax	1440	48	1440 ÷ 48 = 30	25	30 ÷ 25 × 60 = 72
Manicure	1080	72	1080 ÷ 72 = 15	20	15 ÷ 20 × 60 = 45
Mini-facial	1100	44	1100 ÷ 44 = 25	30	25 ÷ 30 × 60 = 50

The eyebrow shape treatment raises most money per hour for the salon, even though the fewest clients take that treatment.

b The salon could increase its income by advertising the eyebrow shape treatment in order to increase the number of clients. The salon could also increase the price of the manicure in order to take advantage of the larger number of clients for this treatment.

Exercise 9B

1 This composite bar chart shows the average daily maximum temperatures for Scotland and Turkey over a 5-month period.

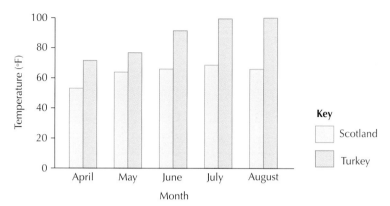

a In which month was the difference in temperature between Scotland and Turkey greatest?

b Describe the trends of the temperatures in Scotland and in Turkey over the 5-month period.

c If this trend in temperatures continues, predict the temperatures in both countries in September.

2 The diagram shows the maximum and minimum temperatures for one day in August in five cities.

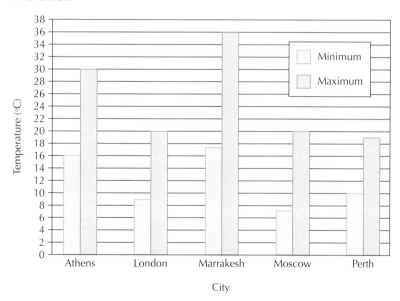

a Which city had the greatest difference between the maximum and minimum temperatures?

b Which city recorded the lowest temperature?

c David was analysing the bar graph and concluded that the maximum temperature is always approximately double the minimum temperature.

Is David correct? Fully explain your answer.

★ **3**   A group of friends decide to enter a 2 km charity 'fun run'.

They record their times before they start training for the event and then after 6 weeks of training.

The back-to-back stem and leaf diagram shows the results.

Before training							After training					
					1	6	7	9				
7	5	4	2	2	2	4	5	6	7	8	8	
9	9	8	7	3	1	3	4	5	7	8		
	8	6	5	2	4	1	5					

Key 1 | 3 represents 31 minutes   Key 1 | 5 represents 15 minutes

a   How many people were in the group of friends who entered the fun run?

b   For the times before training, find:

  i   the median

  ii the lower quartile

  iii the upper quartile.

c   Create boxplots showing the times for the group before and after training.

d   Was the training effective? Use your boxplots to support your answer.

4   The graph shows the sales of whole milk and semi-skimmed milk in a supermarket from 2000 until 2011.

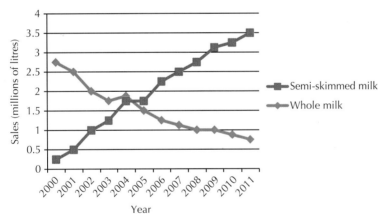

a   How much semi-skimmed milk was sold in 2005?

b   Describe the trend of the sales of both kinds of milk.

c   If these trends continue, predict sales of whole milk and semi-skimmed milk in 2014. Show all your working.

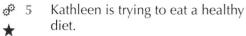 5 Kathleen is trying to eat a healthy diet.

She buys a roast chicken salad sandwich for her lunch. She sees this label on the sandwich box giving the nutritional information about the sandwich.

The pack also has a bar chart showing the recommended daily amounts for each of these nutritional elements.

Kathleen estimates that this sandwich will account for $\frac{1}{4}$ of her nutrient intake for the day.

Is this sandwich a healthy choice for Kathleen? Explain your answer.

Nutrition	Typical values	
	Per 100 g	Per pack (approx 215 g)
Energy value		
kJ	720 kJ	1550 kJ
kcal	170 kcal	365 kcal
Fat	3.8 g	8.2 g
(of which saturates)	0.9 g	1.9 g
Carbohydrate	21.4 g	46.0 g
(of which sugars)	1.3 g	2.8 g
Fibre	1.9 g	4.1 g
Protein	11.7 g	25.2 g
Salt	0.7 g	1.6 g

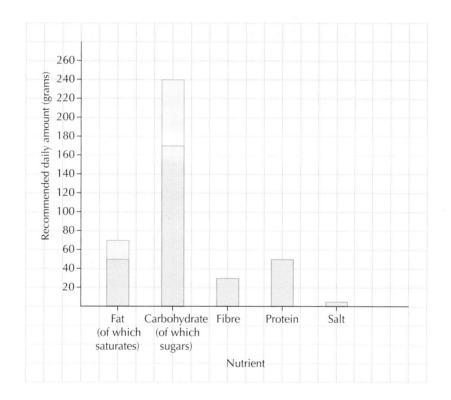

I can represent data graphically. ★ Exercise 9A Q2, Q4

• I can read and interpret data presented graphically. ★ Exercise 9B Q3, Q5

For further assessment opportunities, see the Case Studies for Unit 1 on pages 136–140.

Case studies

Very enterprising

> **This case study focuses on the following skills:**
>
> - using a budget to plan an event **(Ch. 1)**
> - balancing complicated incomings and outgoings **(Ch. 1)**
> - using a combination of statistics to investigate risk and its impact on life **(Ch. 6).**

The Enterprise Group in a school decide to sell decorated candles with holders. They will buy:

- pillar candles that cost £17.50 for a pack of 8
- candle holders that cost £11.50 for a pack of 10
- packs of candle-decorating sets that cost £16.25 each. Each pack is estimated to decorate 50 candles.

The Enterprise Group will put together 200 packs of decorated candles and candle-holders.

They are going to sell them for £4.50 each.

They decide to put a label into each pack warning of the risks of fire due to lit candles.

Their research shows that the probability of a house fire as a result of candle mis-use is 0.08%.

1 a If they make 200 and sell them all, how much profit will they make?   **[2P]**

    b They sell only 140 so decide to sell the remainder of the stock at a reduced price. How much should each remaining candle be sold for in order to break even (assuming that all remaining candles are sold)?   **[1P, 1C]**

                                                               4 marks

2 a How many candles might be used before one caused a house fire to start?   **[1P]**

    b What advice could they put on the label to protect their buyers from house fires? Give two suggestions.   **[1C]**

                                                               2 marks

                                                               Total 6 marks

Wedding dresses

This case study focuses on the following skills:

- using statistics to analyse and compare data sets **(Ch. 7)**
- investigating the impact of interest rates on credit cards and store cards **(Ch. 5)**.

The bridal shop Perfect Days sells wedding dresses.

They offer the following finance package:

```
10% discount on the price of the dress
2.4% interest per month on the balance
```

A credit card company offers 1.9% monthly interest on a balance.

The following boxplot represents a sample of the prices of 20 dresses in Perfect Days and 20 dresses in another bridal shop called Wedding Belles.

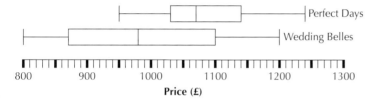

Price (£)

1   a   Stacey wants to buy a wedding dress that costs £1150 from Perfect Days.
If she buys the dress using her credit card and makes no payments, calculate
the total amount to be repaid at the end of 6 months if interest is charged.   **[1S, 1P]**

   b   If she chooses the finance deal offered by the shop, is this a better deal than
the credit card over 6 months?

   Use your calculations to justify your answer.   **[1P, 1C]**

                  4 marks

2   Stacey looked at dresses at both Perfect Days and Wedding Belles. Make two valid
comparisons between the prices of dresses in the two shops.   **[2C]**

                  2 marks

                  Total 6 marks

Frank's case study

> **This case study focuses on the following skills:**
> - converting between currencies involving the use of at least three currencies in a multi-stage task **(Ch. 4)**
> - using a combination of statistical information presented in different diagrams **(Ch. 6)**
> - using statistics to analyse and compare data sets **(Ch. 7)**.

The DGS is a type of gaming PC. Frank finds that a shop in Scotland is selling the DGS for £1399.99.

Frank discovers that the DGS is also sold in the USA, Germany and Japan.

It is sold in the USA for $1892.02.

It is sold in Germany for €1785.25.

It is sold in Japan for ¥245173.

The table shows the exchange rates between the pound (£) and the other currencies.

Country	Currency	Exchange rate
Germany	euro	£1 = €1.2123
Japan	Japanese yen	£1 = ¥171.253
USA	American dollar	£1 = $1.6415

To ship the DGS from the USA costs $120.

To ship the DGS from Germany costs €60.

To ship the DGS from Japan costs ¥9500.

Frank buys some games for the DGS.

The prices of the first six games he buys are:

£32   £30   £35   £12   £20   £27

1 a In which country is the DGS sold at the lowest price – Germany, Japan or the USA? Give a reason for your answer. **[2P, 1C]**

   b To buy the DGS from another country, shipping costs have to be paid.

     i If Frank buys the DGS and pays for it to be shipped to the UK, which is the cheapest option?

     ii Is this cheaper than buying the DGS in the UK? Justify your answer. **[1P, 2C]**

                                                                 6 marks

2 a Calculate the mean and standard deviation for the cost of Frank's games for the DGS. **[1P, 1S, 1P]**

   b Jack is a friend of Frank. He owns a games console and he buys the **same** six games as Frank. The mean price of his six games is £35 and the standard deviation is £11.17.

Make two valid comparisons between the cost of games for the DGS and for consoles. **[1C]**

                                                                     4 marks

                                                                  Total 10 marks

Basketball players

This case study focuses on the following skills:

- using a combination of statistical information presented in different diagrams **(Ch. 9)**
- using statistics to analyse and compare data sets **(Ch. 7)**
- constructing and interpreting a tree diagram **(Ch. 6)**.

The table shows the ages and heights of 10 gifted young basketball players in 2009.

Age (years)	11.9	12.4	12.8	13.2	13.5	13.7	14.7	14.4	14.7	15.2
Height (m)	1.52	1.54	1.62	1.61	1.68	1.72	1.79	1.77	1.88	1.85

Exactly 5 years later in 2014 the basketball players' heights were measured again. The scatter graph shows the results.

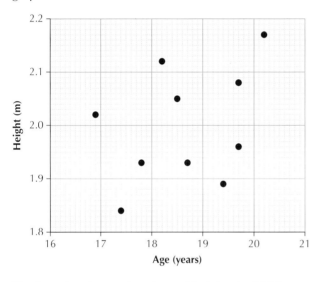

The boxplot shows the players' heights in 2014.

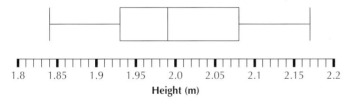

According to the *British Journal of Sports Medicine*, the probability of a player having an ankle injury in a game of basketball is 0.385% and the probability of a player having a head injury in a game of football is 0.78%.

1   a   Plot the 2009 data on a scattergraph.   **[1P]**

      b   Draw a best-fitting line on the graph.   **[1P]**

      c   Henry was a gifted young basketball player in 2009. Estimate his height
           on his 14th birthday.   **[1S]**

      d   Compare the scatter graph you have drawn in part **a** to the scatter graph showing
           the players' heights in 2014. What conclusions might you come to about heights
           and ages of basketball players?   **[1C]**

                                                               4 marks

2   a   For the 2009 data:

           i   calculate the median height

           ii   calculate the interquartile range of the heights

           iii   draw a boxplot to illustrate the heights of the basketball players.   **[1S, 1S, 1P]**

      b   Compare the boxplot given for the 2014 height data and the boxplot
           you have drawn in part a. Make two valid comparisons between their
           heights in 2009 and 2014.   **[1C]**

                                                               4 marks

3   Use the tree diagram below to answer the following questions.

      a   What is the probability of a player getting an ankle injury but not a head injury?

      b   What is the probability of a player getting neither an ankle injury nor a head injury?

      c   Is basketball a risky activity? Give a reason for your answer.

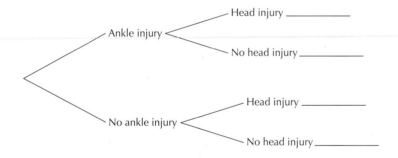

                                                              [1P, 1P, 1C]

                                                                  3 marks

                                                            Total 11 marks

10 Calculating a quantity based on two related pieces of information

This chapter will show you how to:

- calculate a quantity using a formula or relationship involving two or more variables
- solve problems involving the relationships between two or more variables.

You should already know:

- how to calculate a quantity based on a related measurement
- how to calculate a value using a formula involving one variable
- how to solve problems using a simple relationship between variables.

Using formulae

A **formula** is a method of calculating the value of an unknown quantity when you know the values of related quantities.

You need to substitute values for the known quantities into the formula and then calculate the missing amount.

You will need to remember to use the correct order of operations when you evaluate the answer.

You should be able to use many formulae already.

Formulae in geometry

You should already be familiar with these maths and physics formulae.

- Area of a rectangle = length × breadth, $A = LB$
- Area of a triangle = $\frac{1}{2}$ × base × height, $A = \frac{1}{2}bh$
- Pythagoras' theorem, $a^2 + b^2 = c^2$
- The standard deviation of a data sample, $s = \sqrt{\dfrac{\sum(x - \bar{x})^2}{n - 1}}$ and $s = \sqrt{\dfrac{\sum x^2 - \dfrac{(\sum x)^2}{n}}{n - 1}}$
- Distance/speed/time formulae, $D = ST$, $T = \frac{D}{S}$ and $S = \frac{D}{T}$

Other formulae

Formulae are also used to calculate quantities related to everyday real-life problems. These include cooking temperatures, body mass index (BMI), population density and home decoration. Examples of these are covered later in this chapter.

Example 10.1

You can use the formula $A = \pi r^2$ to find the area of a circle of given radius r.

a Work out the area of a circle with diameter 8 cm.

b Work out the radius of a circle which has an area of 113.09 m².

Give your answers to 1 decimal place.

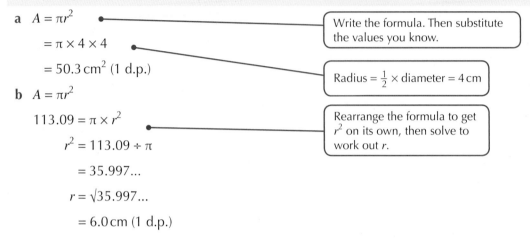

a $A = \pi r^2$

$\quad = \pi \times 4 \times 4$

$\quad = 50.3\text{ cm}^2$ (1 d.p.)

Write the formula. Then substitute the values you know.

Radius $= \frac{1}{2} \times$ diameter $= 4$ cm

b $A = \pi r^2$

$\quad 113.09 = \pi \times r^2$

$\quad\quad r^2 = 113.09 \div \pi$

Rearrange the formula to get r^2 on its own, then solve to work out r.

$\quad\quad\quad = 35.997...$

$\quad\quad r = \sqrt{35.997...}$

$\quad\quad\quad = 6.0\text{ cm}$ (1 d.p.)

Example 10.2

In physics, you can work out the displacement (the distance it has moved), s, of an object using this formula:

$s = ut + \frac{1}{2}at^2$

where u is the initial velocity (speed), a is the constant acceleration and t is the time taken.

A car has an initial velocity of 5 m/s. It accelerates at a constant rate of 2 m/s² for 20 seconds.

How far will it have travelled in this time?

$s = ut + \frac{1}{2}at^2$

$u = 5$ m/s, $a = 2$ m/s², $t = 20$ seconds

$s = 5 \times 20 + \frac{1}{2} \times 2 \times 20^2$

$\quad = 100 + \frac{1}{2} \times 2 \times 400$

$\quad = 100 + 400$

$\quad = 500\text{ m}$

Remember to include units.

Compound measures

Compound measures involve measurements of two different types, such as density (mass and volume), speed (distance and time), and pressure (force and area).

Density (ρ) is the **mass** of a substance per unit **volume**.

This can be expressed in grams per cm^3 or kilograms per m^3.

The relationship between the three quantities is given by the formula

$$\textbf{Density} = \frac{\textbf{Mass}}{\textbf{Volume}}$$

This formula can be arranged to give mass = density × volume and volume = $\dfrac{mass}{density}$

So: $\rho = \dfrac{m}{v}$ $m = \rho \times v$ $v = \dfrac{m}{\rho}$

You can remember the relationships using a triangle.

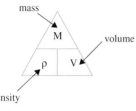

Cover the quantity that you want to find. What you are left with tells you the relationship between the other two quantities (and whether to multiply or divide).

You can also use a formula triangle to work out unknown quantities in problems involving **speed** and **pressure**.

$$s = \frac{d}{t} \qquad\qquad\qquad P = \frac{F}{A}$$

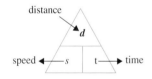   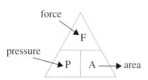

Example 10.3

A slab of concrete has a volume of $60\,cm^3$ and a mass of $150\,g$.

Calculate the density of the concrete.

$\rho = \dfrac{m}{v}$

$= \dfrac{150}{60}$

$= 2.5\,g/cm^3$

Exercise 10A

1   The final speed, v m/s, of a car accelerating at a constant rate is given by the formula $v = u + at$, where u = initial speed in m/s, a = acceleration in m/s^2 and t = time taken in seconds.

Find v when $u = 15\,m/s$, $a = 0.2\,m/s^2$ and $t = 30\,s$.

2   The sum of the squares of the integers from 1 to n is given by the formula

$S = \frac{1}{6}n(n + 1)(2n + 1)$

Use the formula to find the sum of the square of the first 12 integers.

3   The area, A, of a trapezium is given by the formula $A = \frac{1}{2}(a + b)h$ where a and b are the parallel sides and h is the perpendicular height.

Find the area of a trapezium with $a = 6\,cm$, $b = 9\,cm$ and $h = 3\,cm$.

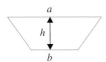

4   Calculate the density of an object with:

   a   mass 45 g and volume 15 cm^3

   b   volume 2.4 m^3 and mass 1812 kg

5   A Formula 1 car completes the Monaco Grand Prix in a time of 1 hour 15 minutes. The race is 260 kilometres in distance.

Calculate the average speed of the car for the whole race. Give your answer in km per hour.

6   The pressure in a steam boiler is 20 000 N/m^2.

The area of one end of the boiler is 0.6 m^2. What is the force of the steam at that end?

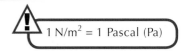
1 N/m^2 = 1 Pascal (Pa)

★ 7   Two statues look identical and both appear to be made out of gold. One of them is a fake.

The density of gold is 19.3 g/cm^3.

Both statues have volume 200 cm^3.

The first statue has mass 5.2 kg.

The second statue has mass 3.8 kg.

Which statue is fake? Use your calculations to explain your answer.

8   Keenan makes candles in the shape of a cylinder.

Each cylinder has a radius of 3 cm and a height of 12 cm.

Keenan buys the candle wax in 10 litre tubs.

How many candles will he be able to make from 1 tub of wax?

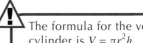
The formula for the volume of a cylinder is $V = \pi r^2 h$.

Remember that 1 cm^3 = 1 ml and 1000 ml = 1 litre.

 Activity

Fergus runs a charity shop. The items listed in the first table have been donated to his shop. Fergus has estimated their volume in cubic centimetres. Now he needs to price them.

The second table provides data about the density of metal and the cost per gram.

Use the information given to work out the value of each item. What price should Fergus put on each item and why?

Item	Volume (cm³)
candlestick (brass)	6
statue (cast iron)	15
ring (gold)	0.5
tankard (stainless steel)	4
jug (silver)	3
plate (copper)	7

Metal	Density (g/cm³)	Cost per gram
brass	8.5	29p
copper	8.9	22p
gold	19.3	£16.58
silver	10.4	80p
stainless steel	7.5	33p
cast iron	7.2	83p

Using formulae in real-life contexts

Many everyday activities involve the use of formulae. For example, every time we bake a cake, convert between temperature in Fahrenheit and Celsius, or convert between miles and kilometres we use a formula. When we use a formula to solve a problem, we often use the ratio between the different variables.

Recipes

Recipes in books or magazines are written to serve a certain number of people, often 4 or 6. If you plan to follow the recipe but want to serve a different number of people, you will need to change the quantity of ingredients accordingly.

The ingredients in a recipe are given in a specified ratio. So, if you need to change the quantities to make a different number of servings, whether you scale up or down, you need to keep the ingredients in the same ratio as in the original recipe. If you change the ratio, the dish may not taste good when it is cooked.

Body mass index

One way of determining if an adult is within a healthy weight range is the body mass index (BMI). This measurement depends on a person's height in metres, h, and mass in kg, w, and uses the formula BMI $= \frac{w}{h^2}$.

The table shows the range of BMI values and associated weight descriptions.

BMI is only a guide to a person's overall health. Many other factors also influence an individual's health, such as the amount of exercise they do, whether they smoke or drink, and whether they eat a balanced diet.

BMI	Weight status
less than 18.5	underweight
18.5–24.9	healthy
25–30	overweight
greater than 30	obese

Stopping distances

The stopping distances for cars can be calculated using a formula. The calculation is based on:

- the driver's thinking distance, which is the distance the car has travelled before the driver reacts to a hazard

- the braking distance, which is the distance the car has travelled once the brakes have been applied.

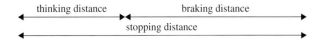

Thinking distance can be increased by tiredness, drink or drugs (legal or illegal), poor eye sight, age and lack of concentration.

Braking distance can be increased by the weight of the car, the condition of the brakes, the condition of the tyres and road conditions.

The formula for calculating stopping distance in good conditions is:

$$\text{stopping distance} = \text{thinking distance} + \frac{\text{speed}^2}{20}$$

where stopping and thinking distances are measured in feet and speed is measured in mph.

In wet road conditions, multiply stopping distance by 2.

In icy road conditions, multiply stopping distance by 10.

⚠ To convert feet into metres, multiply by 0.3048.

The table shows the stopping distance for a typical car at different speeds, when it is being driven in good conditions, that is, in good weather and on dry, well maintained, roads.

Speed (mph)	Thinking distance (ft)	Braking distance (ft)	Stopping distance (ft)
20	20	20	40
30	30	45	75
40	40	80	120
50	50	125	175
60	60	180	240
70	70	245	315

Population density

Density can also be used as a measure of population. The population density of a country or city is the population per square kilometre or square mile. It is a measure of how crowded an area is. The formula for population density is: $\text{population density} = \dfrac{\text{population}}{\text{area}}$

Exercise 10B

1 Here are the ingredients for a peanut butter cheesecake.

For the base	For the filling	To decorate
50 g butter	5 gelatine leaves	270 ml pot double cream
175 g pack peanut cookies	500 g tub ricotta	2 tbsp soft brown sugar
	175 g smooth peanut butter	1 bar peanut brittle, crushed
	175 g golden syrup	
	150 ml milk	

   a   David makes the cheesecake using 150 g of butter. How much peanut butter and milk does he need to make the cheesecake?

   b   The ricotta is on special offer at '600 g for the price of 500 g'. If David uses 600 g of ricotta, how much of each ingredient for the base and filling does he need to make the cheesecake?

★ **2**   These are the ingredients for a strawberry meringue roulade.

For the meringue	For the filling	Icing sugar, for dusting
5 egg whites	150 ml double cream	
275 g caster sugar	200 ml Greek yoghurt	
20 g flaked almonds	225 g strawberries, hulled	

   a   If Marta uses 10 eggs, how much Greek yoghurt does she need to make this roulade?

   b   If she uses 3 eggs, how much double cream does she need to make this roulade?

   c   Marta has 6 eggs and wants to use them all to make a roulade. She also has 250 ml of Greek yoghurt.

      Does Marta have enough Greek yoghurt for a roulade that uses 6 eggs?

3   These ingredients will make 18 chocolate chip cookies.

250 g plain flour	$\frac{1}{2}$ teaspoon bicarbonate of soda
170 g unsalted butter, melted	$\frac{1}{2}$ teaspoon salt
200 g dark brown sugar	1 tablespoon vanilla extract
100 g caster sugar	1 egg
325 g chocolate chips	1 egg yolk

Juan has 300 g of dark brown sugar and 200 g of caster sugar.

   a   Find the maximum number of chocolate chip cookies that he can make using the recipe above.

   b   How much butter will he need?

4   Calculate the BMI and give the weight status of:

   a   Amy, who weighs 54 kg and is 1.52 m tall

   b   Oliver, who weighs 45 kg and is 1.7 m tall.

5   Shona is 65 inches tall and weighs 172 pounds.

   Is Shona a healthy weight?

1 inch = 2.54 cm
1 kg = 2.205 pounds

6   Alan is 1.83 m tall and weighs 61 kg. How much weight does he need to gain to reach a healthy weight?

7   Calculate the stopping distance, in feet, for a car travelling in good conditions at

   a 25 mph   **b** 36 mph   **c** 78 mph

⚙ **8**   Calculate the stopping distance, in metres, for a car travelling in wet conditions at

   a 15 mph   **b** 52 mph   **c** 63 mph

9 Rebecca is driving her car on a motorway at 48 mph in icy road conditions. She sees a warning sign that there has been an accident 500 m ahead.

Will she be able to stop her car before she reaches the accident?

10 The table shows the 10 most populous countries of the world. More than half of the world's population live in these 10 countries.

Calculate the population density for each country. Give your answers to the nearest whole number.

Country	Population	Area (km²)
China	1 362 420 000	9 596 960
India	1 239 500 000	3 287 590
United States	317 546 000	9 629 091
Indonesia	247 008 052	1 919 440
Brazil	201 032 714	8 511 965
Pakistan	185 457 000	803 940
Nigeria	173 615 000	923 768
Bangladesh	149 772 364	144 000
Russia	143 600 000	17 075 200
Japan	127 300 000	377 835

11 The area of Trinidad and Tobago is 1980 square miles. The population density is 460 people per square mile. Work out the population of Trinidad and Tobago.

12 The population of Scotland is approximately 5.2 million and Scotland's population density is 64 people per square kilometre. Work out the approximate area of Scotland.

Use the following information to answer Questions 13–16.

Paint (2 coats of paint needed)

Paint type	Coverage per litre	Price
emulsion	12 m²	£8.95 per 2.5 litre can £16.95 per 5 litre can
gloss	16 m²	£14.98 per 750m*l*

Area to be painted = distance around skirting board of room × height of room, or $A = D \times H$

Wallpaper

Number of rolls of wallpaper needed = $(D \times H) \div 10$

Carpets

Carpets are available in 4 m or 5 m widths.

To work out the amount of carpet needed, measure the longest and widest points of your room, ensuring that you measure into any alcoves and doorways.

Add 10 cm onto each measurement to allow for cutting and trimming.

Carpet is then bought by length, for example, a 2.1 m length of 4 m wide carpet.

★ **13** Three walls of a room with the dimensions shown are to be painted. Calculate the cost of painting the three walls of this room.

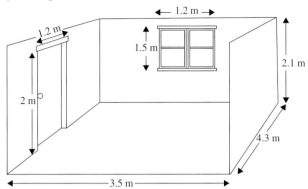

14 The diagram shows a sketch of Peter's dining room. The room is 2.9 m high.

   a Calculate the number of rolls of wallpaper needed to decorate this room.

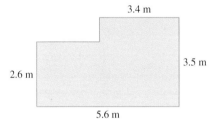

   b If each roll of wallpaper costs £11.99, calculate the cost of the wallpaper.

15 Choose the best width of carpet, 4 m and 5 m, to cover the floors shown in the table below. Work out how many square metres you will need to buy and the cost. The carpet will cost £31.99 per square metre.

Room	Length	Breadth
1	1.8 m	3.6 m
2	2.2 m	3.8 m
3	3.7 m	2.6 m
4	5.2 m	4 m

16 A room measuring 3.2 m by 2.6 m is to be tiled using square tiles of length 30 cm.

   a If the tiles cost £11.99 for a pack of 12, calculate the total cost of the tiles.

   b The carpet fitter can lay the tiles at a rate of 30 tiles per hour.

   She charges £22 plus VAT (charged at 20%) for each hour that she works.

   How much will she charge to lay the tiles?

GO! Activity

Imagine you are a flooring specialist.

Mrs Patel asks you to quote for laying a new floor in her dining room.

The room is 5.3 m long and 4.5 m wide.

She would prefer luxury carpet or wood, but will consider other types of flooring that you recommend.

The prices you charge for various types of flooring are given in the table below, along with other useful information.

Flooring type	Size	Pack size (minimum quantity)	Price	Labour cost to lay the flooring
Carpet tiles	500 mm × 500 mm	Packs of 10	£18.99 per pack	£3.50 per m²
Plain carpet	5 m wide roll	1 m units	£32.95 per 1 m length	£4.00 per m²
Luxury carpet	5 m wide roll	1 m units	£54.99 per 1 m length	£4.00 per m²
Wood (beech)	1.8 m lengths, 200 mm wide	Packs of 5	£28.95 per pack	£8.50 per m²
Wood (oak)	1.8 m lengths, 200 mm wide	Packs of 5	£32.50 per pack	£8.50 per m²
Ceramic tiles	200 mm × 300 mm	Packs of 10	£9.95 per pack	£6.50 per m²

1 Calculate how much it would cost Mrs Patel for each type of flooring. You will need to consider:

   • the amount of flooring material needed, bearing in mind the minimum purchase quantities

   • the cost of the flooring material

   • the labour cost to lay the floor.

2 Mrs Patel has a budget of £450. Can she afford one of her preferred flooring types?

3 What are the advantages and disadvantages of each flooring type? Consider durability, cost, cleaning and any other properties you can think of.

4 Which flooring would you recommend to Mrs Patel? Why would you recommend this type of flooring?

• I can calculate a quantity using a formula or relationship involving two or more variables. ★ Exercise 10A Q7

• I can solve problems involving the relationships between two or more variables. ★ Exercise 10B Q2, Q13

For further assessment opportunities, see the Case Studies for Unit 2 on pages 255–257.

11 Construct a scale drawing, including choosing a scale

This chapter will show you how to:

- enlarge and reduce simple shapes on a grid
- interpret scale diagrams
- construct or sketch diagrams by choosing a scale
- explain decisions based on the results of measurements.

You should already know:

- how to convert between metric units
- how to read scales
- how to use measuring instruments accurately
- how to apply scales.

Enlarging and reducing simple shapes

In this section, you will enlarge or reduce a shape when you are given the **scale factor** (SF).

For an enlargement, the scale factor will be a number greater than 1.

For a reduction, the scale factor will be a number less than 1, that is, a fraction.

Example 11.1

a Enlarge each of these shapes using a scale factor of 2.

b Reduce each of these shapes using a scale factor of $\frac{1}{3}$.

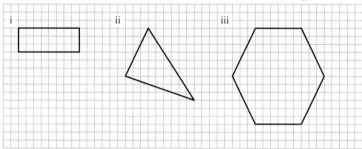

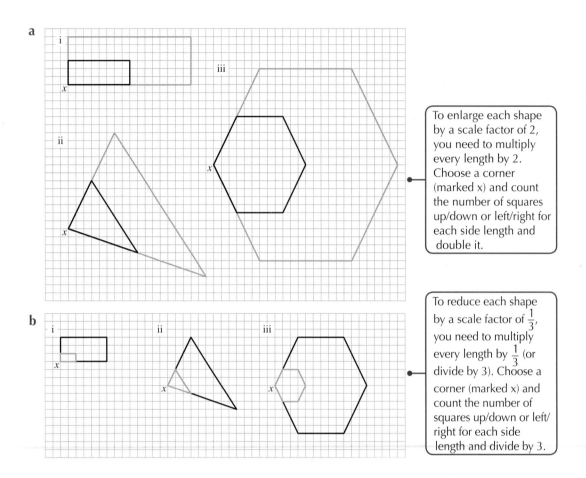

To enlarge each shape by a scale factor of 2, you need to multiply every length by 2. Choose a corner (marked x) and count the number of squares up/down or left/right for each side length and double it.

To reduce each shape by a scale factor of $\frac{1}{3}$, you need to multiply every length by $\frac{1}{3}$ (or divide by 3). Choose a corner (marked x) and count the number of squares up/down or left/right for each side length and divide by 3.

Exercise 11A

★ 1 Copy each shape from the diagram below onto squared paper. Enlarge each shape by:

   a   scale factor 2

   b   scale factor $\frac{1}{2}$

   c   scale factor $3\frac{1}{2}$

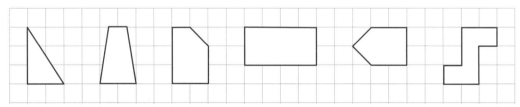

Scale drawings

A **scale drawing** is an accurate representation of a real object.

On most occasions the scale drawing will be smaller than the real object, although electronic parts, circuits, watch movements and cells will require an enlargement. The object is enlarged or reduced by a given scale factor.

Scale drawings are often used in everyday life. Examples include:

- construction industry: plans of motor vehicle parts, plans for a ship
- planning department: plan for the construction or extension of a house, plan for development of a piece of ground
- maps: road maps, ordinance survey maps
- models: model kits, replicas of vehicles.

If you are not given the diagram on squared paper, you will need to use a ruler and a protractor to construct your drawing.

The angles of a shape will not change when you enlarge or reduce the shape; only the lengths of the sides will change.

Interpreting scale diagrams

The scale on a diagram can be written in two ways – as a ratio or as a representative fraction.

Scale as a ratio

Maps often use a **ratio** to display scale. The ratio often includes units of measurement.

Examples of this type of scale include: $1\,cm:5\,m$, 4 inches : 1 mile, $2.5\,cm:5$ miles.

Example 11.2

A plan of a garden uses a scale of $1\,cm:5.5\,m$.

The length of the garden is $9.7\,cm$ on the plan.

What is the actual length of the garden?

$$9.7 \times 5.5 = 53.4\,m$$ ●————— (Each $1\,cm$ on the plan is equivalent to $5.5\,m$ in real life.)

Scale as a representative fraction

On maps scales are sometimes written as **representative fractions** (RF).

If a scale on a map is given as $\dfrac{1}{800\,000}$, this is the same as the ratio $1:800\,000$.

This means $1\,cm$ represents $800\,000\,cm$, so $1\,cm$ represents $8000\,m$ and $1\,cm$ represents $8\,km$.

Other examples of this type of scale include: $\dfrac{1}{25}$, $\dfrac{1}{30\,000}$, and $\dfrac{3}{10\,000}$

Notice that no units of measurement are given in the fraction form.

A scale of $\dfrac{1}{25}$ could mean $1\,mm:25\,mm$ or $1\,cm:25\,cm$ or $1\,m:25\,m$, etc.

Example 11.3

A map has a scale of $\dfrac{1}{250}$

a A path shown on the map is 7.2 cm long.

How long is the path in real life? Give your answer in metres.

b The distance between Joe's house and Laura's house is 35 metres in real life.

What will this distance be on the map? Give your answer in centimetres.

a

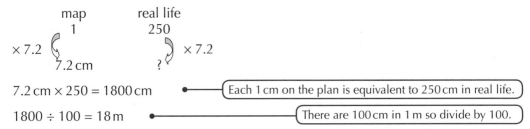

map   real life

× 7.2   1   250   × 7.2

7.2 cm   ?

7.2 cm × 250 = 1800 cm   Each 1 cm on the plan is equivalent to 250 cm in real life.

1800 ÷ 100 = 18 m   There are 100 cm in 1 m so divide by 100.

The path will be 18 m long in real life.

b 35 m ÷ 250 = 0.14 m

0.14 m × 100 = 14 cm

The distance between the houses on the map will be 14 cm.

Exercise 11B

1   **a**   Draw this shape accurately.

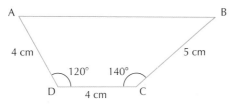

     b   Use your drawing to find the length of the side AB.

2   **a**   Draw this shape accurately.

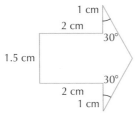

     b   Enlarge this shape using a scale factor of 3.

     c   Use your enlargement to find the overall length of the enlarged arrow.

3 Make an accurate drawing of each diagram.

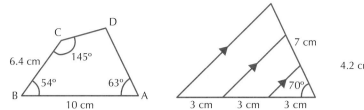

4 Accurately draw an equilateral triangle of side 6 cm. What is the height of the triangle?

5 A triangle ABC has ∠ABC = 40°, AB = 7 cm and AC = 5 cm. There are two different triangles that can be drawn using this information. Accurately draw and label both triangles.

★ **6** The map shows part of the county of Kent in the south of England.

   a Jack lives in Hastings and works for a delivery firm.

   He travels from Hastings to Dover and then to Canterbury before returning to Hastings.

   What distance has he travelled?

   b Sheila lives in Ramsgate and visits her cousin in Maidstone. What distance does she travel from Ramsgate to Maidstone?

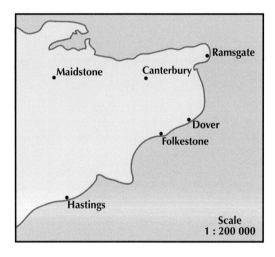

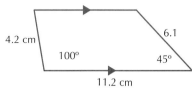

Measure the distance 'as the crow flies' in each case.

7 The map shows a boat journey around an island. If the boat has a top speed of 12 km/h, how long will it take to travel all the way around the island?

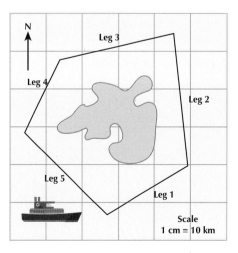

8 A plan of Edinburgh Castle is shown below.

A tour guide estimates her average walking speed on a tour as 15 metres per minute.

Her tour starts at 1240 at the ticket office and visits the Portcullis gate, cartsheds, military prison, reservoirs and hospital.

Will she be finished before 1315? Give a reason for your answer.

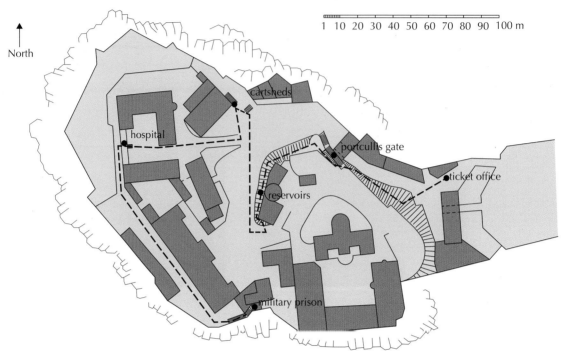

9 The diagram shows the front of a kennel drawn to a scale of $\frac{1}{30}$

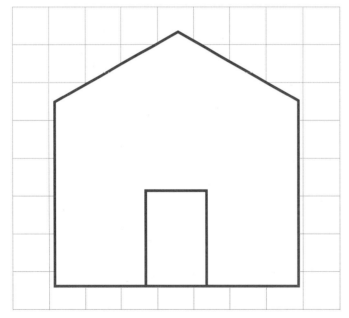

a What are the dimensions of the door opening?

b What is the overall height of the kennel?

c The front of the kennel is to be painted. What is the area that will be need to be painted?

10 The diagram is a scale drawing of the Great Beijing Wheel in China.

The height of the wheel is 210 m.

What scale has been used for this diagram?

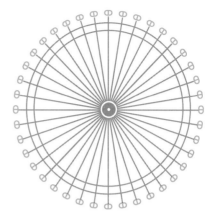

11 The scale drawing shows the route a ferry travels to get from Port A to Port B.

   a What scale has been used for this diagram?

   b What is the actual distance between port A and port B?

★ 12 The map shows the Cairngorms National Park

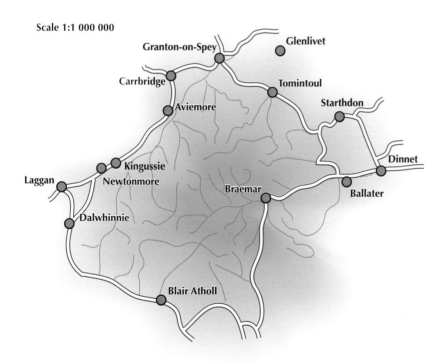

Scale 1:1 000 000

   a As the crow flies, what is the distance in km from:
   i   Braemar to Ballater
   ii Tomintoul to Dalwhinnie?

b **i** A hiker walks at an average speed of 3 km/h. How long will it take her to walk from Dalwhinnie to Granton-onSpey, if she keeps to paths and roads?

ii The hiker always starts hiking at 10:00 and makes camp after walking for 8 hours. When will she arrive in Glenlivet?

13 Here is part of a map of Scotand.

a What is the distance as the crow flies from:

i Aberdeen to Glasgow

ii Oban to Dundee

iii Inverness to Stirling?

b A Eurofighter jet plane can travel at 2500 km/h and has a range of 3600 km. Can the jet start at Inverness, fly over Aberdeen, Edinburgh, Glasgow and Oban, and return to Inverness without refuelling? Give a reason for your answer.

c How long would it take a Eurofighter jet to fly from Edinburgh to Aberdeen?

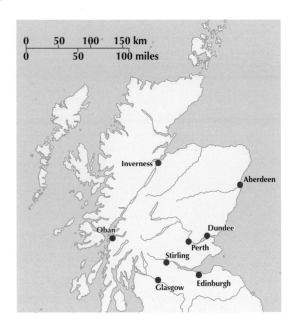

Constructing scale drawings

This section shows you how to interpret a sketch so that you can produce an accurate scale drawing.

Example 11.4

The diagram shows a ladder leaning against a wall.

The bottom of the ladder is 1.5 m from the foot of the wall.

The ladder reaches 3.5 m up the wall.

a Using a scale of 1:50, make a scale drawing to show the ladder and the wall.

b What is the actual length of the ladder?

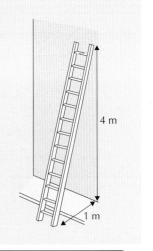

a Scale 1:50 ●———————————————— (Step 1: write down the scale.)

distance from the foot of the wall = 150 cm ÷ 50 = 3 cm

distance up the wall = 350 cm ÷ 50 = 7 cm

(Step 2: work out the lengths of the lines to be drawn. Convert measurements in metres to cm.)

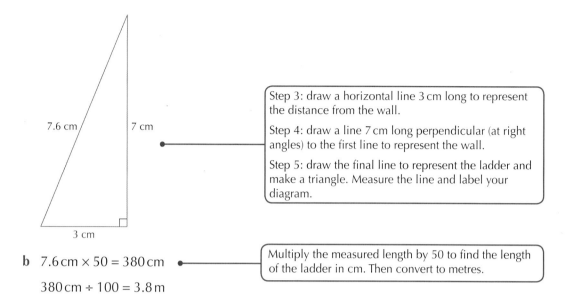

Step 3: draw a horizontal line 3 cm long to represent the distance from the wall.

Step 4: draw a line 7 cm long perpendicular (at right angles) to the first line to represent the wall.

Step 5: draw the final line to represent the ladder and make a triangle. Measure the line and label your diagram.

b $7.6\,\text{cm} \times 50 = 380\,\text{cm}$

$380\,\text{cm} \div 100 = 3.8\,\text{m}$

Multiply the measured length by 50 to find the length of the ladder in cm. Then convert to metres.

Exercise 11C

★ **1** The sketch shows the dimensions of a new play park that is being designed. A fence is to be erected all the way around the park.

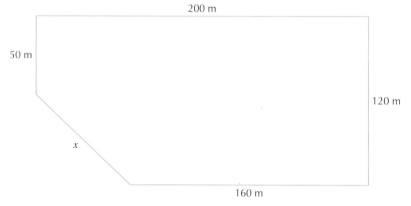

a Use a scale of 1 cm : 20 m to make a scale drawing of the park.

b Use your scale drawing to find the value of x and therefore the length of the fence.

2 This is a sketch of a garden plot.

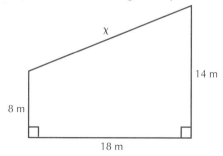

a Make an accurate scale drawing of the garden. Choose a suitable scale.

b Measure your scale drawing to find the length of side x.

3 This sketch shows a car park.

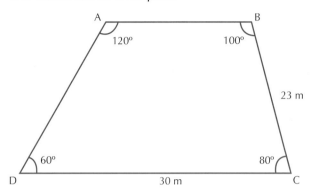

a Make a scale drawing of the car park. Choose a suitable scale.

b What is the length of the longer diagonal in metres?

4 A supply vessel is carrying supplies to a North Sea oil rig.

The vessel leaves Aberdeen harbour and sails 130 km due east. It then sails 60 km due north and arrives at the oil rig at 17:30.

a Draw a sketch of the supply vessel's journey.

b Choose a suitable scale and make an accurate scale drawing of the journey.

c Use your scale drawing to find the shortest distance from the oil rig back to Aberdeen harbour.

d A helicopter can fly at a speed of 207 km/h. At what time should the helicopter leave Aberdeen to arrive at the oil rig at the same time as the supply vessel?

5 Find the height of the tree by choosing a suitable scale and making an accurate scale drawing.

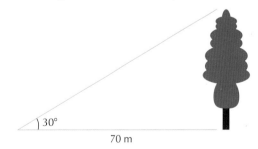

6 Mike is at a viewpoint 130 m above sea level and notices a tall ship passing. The angle of depression is 50°.

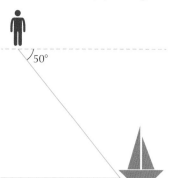

> The angle of depression is the angle measured from the horizontal when you look down at something.

Using a suitable scale, make a scale drawing to find how far the ship is from Mike.

★ **7** The diagram shows a plan of a kitchen.

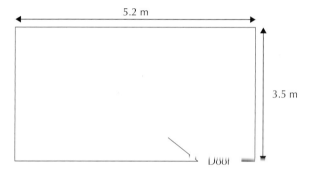

The table below shows the dimensions of various kitchen appliances and units that could be fitted into the kitchen.

Appliance/Unit	Length (cm)	Depth (cm)
Fridge freezer	60	60
Cooker	100	60
Washing machine	60	60
Dish washer	60	60
Tumble dryer	60	60
Sink	95	60
Drawer unit	50	60
Base unit	100	60
Base unit	80	60
Base unit	50	60
Base unit	30	60

a Choose which appliances will be in the kitchen and also how many base units will be required. Give reasons for your choices.

b Make a sketch of the kitchen, marking the positions of the appliances and units.

c Using a suitable scale, make an accurate scale drawing of the kitchen with appliances and units in place.

GO! Activity

The plan below shows the dimensions of a new supermarket.

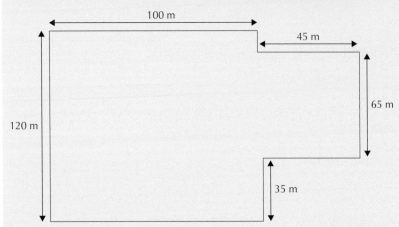

a Decide where the entrance, exit and checkouts will be placed.

b By visiting a local supermarket or looking on the internet, research some types of products sold by a supermarket. Investigate how much shelf space is given to each type of product.

c Make a sketch of the design showing the position of each product.

d Using an appropriate scale, make a scale drawing of the supermarket.

e Why do you think some products are placed where they are? Give some reasons.

- I can complete accurate enlargements and reductions of diagrams using a given scale factor. ★ Exercise 11A Q1

- I can interpret scale diagrams. ★ Exercise 11B Q6, Q12

- I can construct scale diagrams. ★ Exercise 11C Q1

- I can explain decisions based on the results of measurements. ★ Exercise 11C Q7

For further assessment opportunities, see the Case Studies for Unit 2 on pages 255–257.

12 Planning a navigation course

This chapter will show you how to:

- use bearings to create an accurate plan
- create and interpret an accurate plan
- use bearings and the relationships between speed, distance and time to solve navigation problems.

You should already know:

- that bearings are measured from north in a clockwise direction and have three figures
- how to convert between metric units
- how to read and apply scales
- how to use measuring instruments accurately.

Planning a navigation course

Navigation is important for journeys by land, sea and air. To navigate accurately from one position to another both a **distance** and a **bearing** are required. It is crucial that the measurements be accurate for effective planning.

Bearings and back bearings

A bearing is always measured from north in a clockwise direction.

A bearing is described using three figures in order to avoid any confusion between, for example, 045°, 145° and 245°. If the navigator knows all three figures, then there is no confusion. But if they hear only 45°, it is unclear if a bearing of 045°, 145° or 245° was intended. Any bearing that is less than 100° needs to start with a zero.

Example 12.1

Masood travels on a bearing of 075° for 300 metres.

Use a scale of 1 cm : 50 m to show Masood's journey.

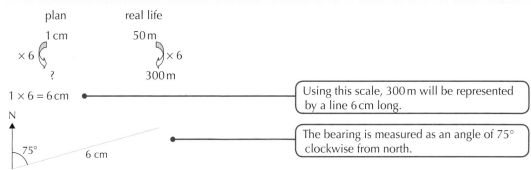

Using this scale, 300 m will be represented by a line 6 cm long.

The bearing is measured as an angle of 75° clockwise from north.

If the bearing is known from point A to point B in a journey, it is often useful to know the bearing required for the return journey, that is, from point B back to point A. This is called the **back bearing** or the **return bearing**.

Example 12.2

Lucy walks from A to B on a bearing of 065°.

What is the back bearing?

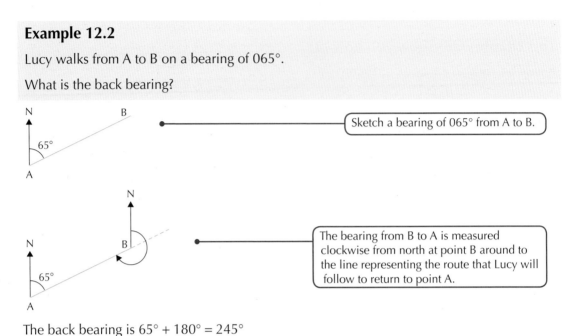

Sketch a bearing of 065° from A to B.

The bearing from B to A is measured clockwise from north at point B around to the line representing the route that Lucy will follow to return to point A.

The back bearing is 65° + 180° = 245°

Scale drawings and bearings

You can use bearings in conjunction with scale drawings to help you work out distances in real life.

Example 12.3

A ship sails 25 km from port A on a bearing of 070° to port B.

It then sails 30 km due east to port C, before returning directly to port A.

Using a suitable scale, make a scale drawing to find the distance and bearing of the return journey.

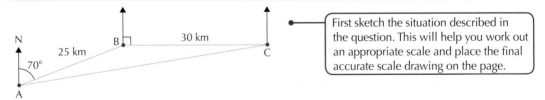

First sketch the situation described in the question. This will help you work out an appropriate scale and place the final accurate scale drawing on the page.

Choose a scale to fit the diagram on the page.

If your diagram is too small, you will not be able to measure lengths accurately in your finished drawing. If it is too big, it might not fit on the page.

A scale of 1 cm : 5 km will give a reasonable size diagram.

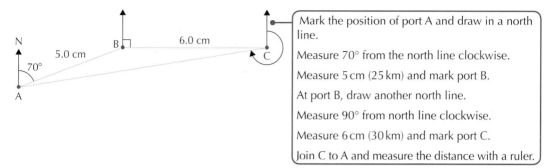

Mark the position of port A and draw in a north line.

Measure 70° from the north line clockwise.

Measure 5 cm (25 km) and mark port B.

At port B, draw another north line.

Measure 90° from north line clockwise.

Measure 6 cm (30 km) and mark port C.

Join C to A and measure the distance with a ruler.

Distance from C to A on scale drawing is 10.8 cm.

Actual distance is 10.8 × 5 = 54 km.

Bearing of return journey (from C to A) is 180° + 81° = 261°.

Measure the angle clockwise from the north line at C as far as the line CA. Do this in two parts, from north to south (180°) and then from south to the line CA (81°).

Exercise 12A

1 The map shows part of the Cairngorm National Park near Aviemore.

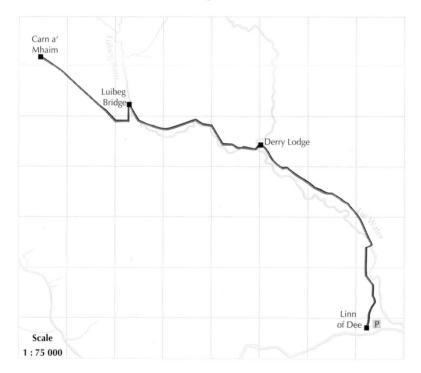

Scale
1 : 75 000

Use the map to find:

a the bearing and distance from Derry Lodge to Linn of Dee

b the return bearing from Linn of Dee to Derry Lodge

c the bearing and distance of Luibeg Bridge to Cairn a' Mhaim.

2 Look at the Ordnance Survey map of Keswick.

OS map of Keswick.
Scale 1 : 50 000

a What is the bearing and distance of Silver Hill from Little Braithwaite?

b What is the bearing and distance of Lord's Island from St Herbert's Island?

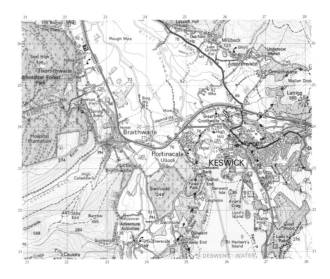

3 Katy is planning her Duke of Edinburgh Award expedition in the Pentland Hills near Edinburgh.

She plans to set off from the Flotterstone Inn and walk to the trout fishery at Loganlea Reservoir. She will then walk to Carlops where she can catch a bus to take her home to Penicuik. Use the map of the area to find the bearing and distance for each part of Katy's expedition.

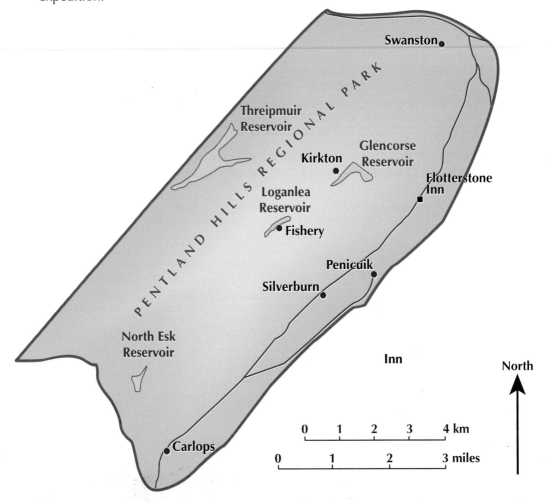

4   James has been designing an orienteering course.

The scale drawing shows the route taken, but the bearings and distances have been omitted.

Write instructions with bearing and distances that can be given to the competitors so that they can complete the course.

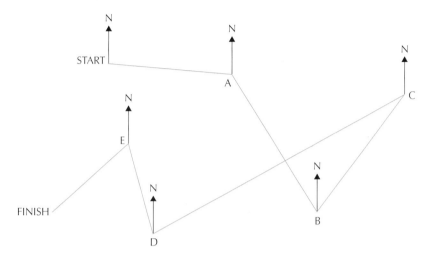

Scale: 1 cm : 120 m

★ 5   Adrian lives in the country side. He leaves his home and jogs 5 km on a bearing of 150°. He then walks 2 km on a bearing of 220°.

a   Draw a sketch showing Adrian's route.

b   Choose a suitable scale and make an accurate scale drawing of his route.

c   How far is Adrian from home after his 2 km walk?

d   What is the bearing that he will have to take for a direct route home?

6 Jemima is watching a sailing regatta. She observes the Flying Duckling on a bearing of 075° and at a distance of 220 m. She also observes the finish marker buoy on a bearing of 140° at a distance 160 m.

    a Choose a suitable scale and make a scale drawing of what Jemima can see.

    b What distance is the Flying Duckling from the finishing buoy?

    c What bearing will the Flying Duckling have to follow in order to get to the finishing buoy in the shortest time?

★ 7 During the Tall Ships Regatta, two ships sail out of port.

The Dolphin sails for 2 hours 30 minutes at a speed of 5 miles per hour on a bearing of 055°, then weighs anchor.

The Star sails for 1 hour 45 minutes at 6 miles per hour on a bearing of 210°, then weighs anchor.

    a Choose a suitable scale and make an accurate scale drawing to show the positions of the two ships.

    b How far apart are the two ships?

    c What is the bearing of the Star from the Dolphin?

★ 8 A helicopter is delivering supplies to three oil rigs. The average speed of the helicopter is 120 mph.

It leaves the airport, point A, and flies for 20 minutes on a bearing of 250° to platform Bravo, point B.

It then flies for 30 minutes on a bearing of 160° to the second oil platform Charlie, point C.

It then flies for 15 minutes on a bearing of 022° to the final oil rig Delta, point D.

Finally it flies directly back to the airport at A.

    a Chose a suitable scale and make a scale drawing to show the journey of the helicopter.

    b What is the distance and bearing of the final stage of the helicopter's return journey from D to A?

★ **9** Mhairi is taking part in an orienteering competition. She is given these instructions.

> ### Orienteering instructions
> * *Start to checkpoint A: 800 m on a bearing of 150°*
> * *Checkpoint A to checkpoint B: 600 m on a bearing of 085°*
> * *Checkpoint B to checkpoint C: 1200 m on a bearing of 340°*
> * *Checkpoint C to checkpoint D: 1000 m on a bearing of 110°*
> * *Return to start*

Mhairi runs at an average speed of 6 km/h.

a Chose a suitable scale and make a scale drawing of Mhairi's route.

b What is the bearing from checkpoint D back to the start?

c How long will it take Mhairi to complete the whole course?

10 The diagram shows a port P and two harbours X and Y on the coast.

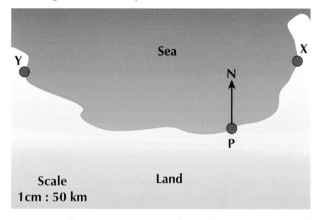

a Write the instructions to show the bearing and distance for a ship that starts at port P and travels to harbours X and Y before returning to port.

b Write the instructions for the reverse of the journey in part **a**.

11 The diagram shows the mouth of an estuary and three small harbours, A, B and C. Davendra is swimming in the estuary and is at point D.

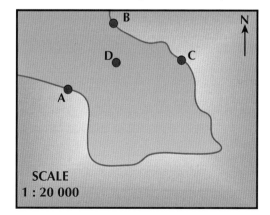

Which of the three harbours is Davendra closest to? Give the bearing and distance of the closest harbour from the swimmer.

12 The diagram shows the control points set up ready for an orienteering course.

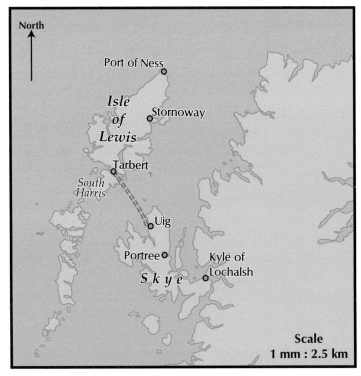

The organisers want to create a different course for each of three different levels of experience: beginners, intermediate, and experts.

The beginner course must be less than 1.5 km, the intermediate course must be longer than the beginner course but less than 2 km, and the expert course must be longer than 2 km. The course can start at any point and must visit every point once.

Create a course for each level. Write the bearing and distance for each leg of your courses and justify why each course meets the criteria.

★ 13 The diagram shows the ferry route from Uig to Tarbert.

a   Choose a suitable scale and use a ruler and protractor make an accurate drawing to show the two ports of Uig and Tarbert.

b   A submarine is observed travelling on the surface, on a bearing of 287° from Uig and 184° from Tarbert. Plot the submarine on your diagram.

c   A container ship is on a bearing of 002° from Uig and 108° from Tarbert. Plot the container ship on your diagram.

d   The submarine is travelling on a bearing of 046° at the same speed as the container ship. The container ship is travelling on a bearing of 210°. The submarine must remain at least 3000 metres from shipping at all times. Will the submarine have to alter course to avoid the container ship? Justify your answer.

🔘 Activity

A sailing race is being planned for an area of water measuring 8 km by 8 km.

Criteria for the race:

- there will be four marker buoys that the yachts must sail around
- each leg of the race must be at least 5 km long
- there must at least one bearing between each of 0°–90°, 90°–180°, 180°–270°, and 270°–360° during the course of the race
- the yachts must sail back to the starting point at the end of the race.

    a   Design a possible course for the race.

    b   Using an appropriate scale, make an accurate scale drawing of the course that you have planned.

- I can create and interpret an accurate plan.
  ★ Exercise 12A Q5

- I can use bearings to create an accurate plan.
  ★ Exercise 12A Q9

- I can use the relationships between speed, distance and time to solve navigation problems.
  ★ Exercise 12A Q7, Q13

For further assessment opportunities, see the Case Studies for Unit 2 on pages 255–257.

13 Carrying out efficient container packing

This chapter will show you how to:

- analyse a problem involving container packing
- use appropriate strategies to find more than one solution
- use the first-fit and the first-fit decreasing algorithms to work out the most efficient container packing.

You should already know:

- how to calculate the quantity of an item that will fit into a given space.

Stacking and filling

A very important skill in everyday life is to be able to pack smaller items into a box, suitcase or cupboard in the most efficient way. This skill will help you keep a tidy house and help when you are packing to go on holiday.

In many real-life situations items will be stacked for storage and transportation purposes. When products are transported they are often stacked on wooden **pallets**.

A pallet is a flat transport structure that supports goods while being lifted by a forklift truck. Goods are secured on a pallet with strapping or shrink-wrap.

Pallets come in different sizes and can be stacked to different heights.

Supermarket packaging

Goods transported to supermarkets usually have three layers of packaging. These are:

- **primary packaging:** this is the wrapping or the containers that are handled by the customer. This needs to contain and protect any food product, as well as provide information and be attractive to the consumer.

- **secondary packaging:** this is the middle layer of packaging that uses larger wrappings, containers or boxes to group quantities of primary packaged goods. The goods might be displayed on the shelf in the secondary packaging.

- **transit packaging:** this is the outer container that uses wooden pallets, plastic and board wrapping to make the transportation easier.

An example of this would be a typical stock cube.

Each is a cube of side 2 cm.

The cubes come in packs of 6, 12 and 24. This is the primary packaging.

These packs are then grouped into trays with film wrap which is the secondary packaging.

Finally, when a supermarket places a large order for the stock cubes, the manufacture will stack the film wrapped packs onto a pallet for delivery.

transit packaging

secondary packaging

primary packaging

Example 13.1

Toy cars are individually packed in small boxes measuring 8 cm by 5 cm by 4 cm.

What is the greatest number of these small boxes that can be packed into a larger container measuring 30 cm by 28 cm by 18 cm?

There are a number of different ways (orientations) in which the small boxes can be packed into the larger container. Each way needs to be considered to find the most efficient arrangement.

Method 1

Place the 8 cm side of the small box along the 20 cm side of the large container, the 5 cm side along the 30 cm side and the 4 cm side along the 18 cm side.

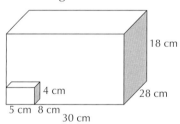

$28 \div 8 = 3.5 \approx 3$ $30 \div 5 = 6$ $18 \div 4 = 4.5 \approx 4$

$3 \times 6 \times 4 = 72$

> Use the lengths of the sides to work out the greatest number of small boxes that can be placed along the three sides. Remember to round down.

Method 1 will fit 72 toy cars.

Method 2

Large container side (cm)	Small box side (cm)	Maximum number of small boxes
28	8	3
30	4	7
18	5	3

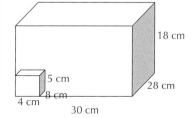

$3 \times 7 \times 3 = 63$ toy cars

(continued)

Method 3

Large container side (cm)	Small box side (cm)	Maximum number of small boxes
30	8	3
28	5	5
18	4	4

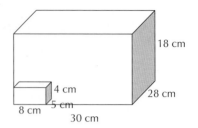

$3 \times 5 \times 4 = 60$ toy cars

Method 4

Large container side (cm)	Small box side (cm)	Maximum number of small boxes
30	8	3
28	4	7
18	5	3

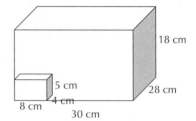

$3 \times 7 \times 3 = 63$ toy cars

Method 5

Large container side (cm)	Small box side (cm)	Maximum number of small boxes
18	8	?
28	5	5
30	4	7

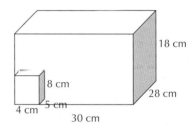

$2 \times 5 \times 7 = 70$ toy cars

Method 6

Large container side (cm)	Small box side (cm)	Maximum number of small boxes
18	8	2
28	4	7
30	5	6

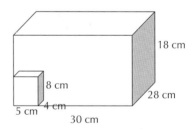

$2 \times 7 \times 6 = 84$ toy cars

Exploring all six methods shows that the maximum number of toy cars that can be packed into the larger container is 84.

Note:

You cannot simply divide the volume of the larger container by the volume of the small box.

The volume of the large container is $28 \times 30 \times 18 = 15\,120\,\text{cm}^3$.

The volume of the small box is $8 \times 5 \times 4 = 160\,\text{cm}^3$.

$15\,120 \div 160 = 94.5$

94 toy cars cannot be packed into the large container as there is some 'spare' space inside the large container that is not completely filled by the toy cars. Dividing the large volume by the small volume does not take into account this extra space and therefore gives an incorrect solution.

Exercise 13A

1 A cylindrical can of soup has diameter 7.8 cm and height 10.5 cm.

The soup cans arrive at the supermarket packed in a cardboard box with dimensions 24 cm by 32 cm by 21.5 cm.

Calculate the maximum number of cans of soup there can be inside this box.

7.8 cm
10.5 cm

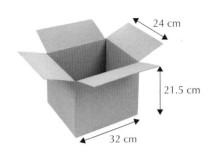
24 cm
21.5 cm
32 cm

2 Some products can be purchased in cylindrical tins or in Tetra Pak cartons, which are in the shape of a cuboid.

The tin of tomatoes has diameter 7 cm and height 10.5 cm.

The Tetra Pak has dimensions 8.4 cm by 4.5 cm by 10.5 cm.

a A chef wants to buy chopped tomatoes for her kitchen.

The tomatoes will be packed into a cardboard box in the shape of a cuboid, with base 45 cm by 50 cm and height 12 cm.

Calculate the maximum number of each type of container of tomatoes that can be packed into this box.

b The tin contains 400 g and the carton contains 390 g.

Which option from part **a** will give the chef more tomatoes?

★ 3 A thermos flask is packaged in polystyrene to ensure that it does not break.

The flask and its polystyrene protection are then packed in a box measuring 10 cm by 12 cm by 15 cm.

Two flask boxes are then packaged together in a larger box.

a Write down three possible sets of dimensions for the larger box.

b Draw a sketch of each of the three possible larger boxes.

c The box which is selected uses the least amount of cardboard. Which size of box is the best choice? Use your calculations to justify your choice.

d A customer wants to buy 24 flasks. Design a suitable container to hold these flasks. Make a sketch of your design showing its dimensions.

4 Gail buys milk in 1 litre cartons to sell in her shop.

a Each carton has a square base with side 7 cm.

What is the minimum height that the carton must be to contain at least 1 litre of milk? Give your answer to the nearest millimetre.

b To be transported to Gail's shop, the cartons will be packed in a cardboard box which has dimensions 30 cm by 35 cm by 28 cm.

Calculate the maximum number of milk cartons that will fit into the cardboard box. The cartons do not need to be packed upright in the cardboard box.

full fat   full fat
MILK   MILK
1 Litre

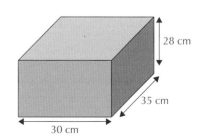
28 cm
35 cm
30 cm

5 Arnie is buying fruit juice for a party.

The fruit juice cartons will be packed in a cardboard box which is 42 cm by 27 cm by 20 cm.

Each juice carton is 6 cm by 9 cm by 19.5 cm.

Calculate the maximum number of cartons of juice that can be packed in the cardboard box.

The cartons do not need to be packed upright.

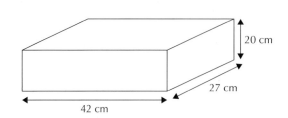

20 cm

27 cm

42 cm

★ **6** Apples are packaged ready for display in a single layer in an open tray which has dimensions 600 mm × 300 mm × 105 mm. The trays must be stacked upright.

These open trays are then stacked onto a pallet measuring 1200 mm × 1000 mm to be transported to the supermarket.

a Calculate the maximum number of open trays which will fit in one layer onto this pallet.

b The open trays can be stacked to a height of 800 mm on the pallet. What is the maximum number of open trays that can be fitted onto a single pallet?

7 Apples can also be packaged in closed boxes which have dimensions 500 mm by 300 mm by 250 mm.

They are then stacked onto a pallet measuring 1000 mm × 1000 mm.

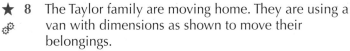

a Calculate the number of boxes which will fit in a single layer onto this pallet.

b The boxes can be stacked to a height of 1200 mm. What is the maximum number of boxes that can be fitted onto this pallet?

★ **8** The Taylor family are moving home. They are using a van with dimensions as shown to move their belongings.

The larger items of furniture have already been moved. The remaining items will be wrapped and put into boxes. The boxes come in three different sizes as shown on the next page.

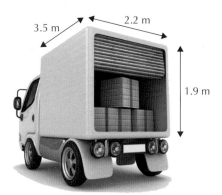

3.5 m   2.2 m

1.9 m

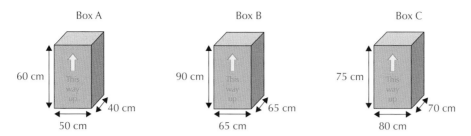

Box A · 60 cm · 50 cm · 40 cm · This way up

Box B · 90 cm · 65 cm · 65 cm · This way up

Box C · 75 cm · 80 cm · 70 cm · This way up

When the Taylors have finished packing, they find that they have filled 28 boxes of size A, 14 boxes of size B and 17 boxes of size C.

No more than two boxes can be stacked on top of each other in the van and they must be stacked the correct way up.

What is the minimum number of trips that the Taylors will need to make to transport all their belongings to their new home? Use your working to justify your answer.

9   In shops, items on special offer are often put at the end of aisles to encourage customers to buy them.

They are stacked on full pallets, half-pallets or quarter-pallets depending on the total weight of the product.

The pallets have dimensions as shown below.

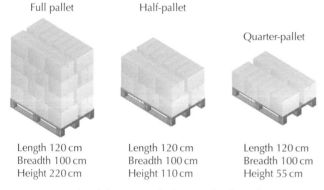

Full pallet
Length 120 cm
Breadth 100 cm
Height 220 cm

Half-pallet
Length 120 cm
Breadth 100 cm
Height 110 cm

Quarter-pallet
Length 120 cm
Breadth 100 cm
Height 55 cm

a   For each of the items below calculate the maximum number which can be stacked onto the given pallet type.

Item	Dimensions ($l \times b \times h$)	Pallet type
Kitchen roll (24 roll multipack)	40 cm × 33 cm × 46 cm	full
Fridge pack of cola (12 cans)	40 cm × 13 cm × 14 cm	quarter
Box of crisps (48 packets per box)	40 cm × 32 cm × 30 cm	half

b   The kitchen roll costs £17.99 per multipack, the fridge pack of cola costs £5.45 and the box of crisps costs £19.99.

What is the total retail value of each pallet when fully stacked?

10 Campbell has a large DVD collection.

He has recently bought a shelving unit to store them on.

The shelves on the unit are 70 cm wide and 21 cm high. The unit has five shelves.

Each DVD is 14 mm wide and 190 mm high.

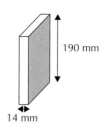

190 mm

14 mm

a What is the maximum number of DVDs that will fit onto the shelving unit?

b Campbell currently has 187 DVDs and he buys one new DVD every month.

How long will it be until the storage unit is full?

c Campbell has also been buying Blu-ray discs which have a width of 12 mm and a height of 170 mm.

What is the maximum number of these which can fit onto one shelf?

11 Olivia transfers books between libraries in her van. The books are packed in open crates and must be stacked upright. The crates are cuboids measuring 60 cm by 30 cm by 20 cm.

The van has a rectangular floor space of 290 cm by 180 cm and the crates cannot be stacked more than five high.

On Thursday Olivia has to transfer 140 crates of books. Will she be able to move all the books in one journey? Use your working to justify your answer.

🔵 Activity

Soft drinks are sold in different sizes of cans, which are available in different sizes of pack.

The most common size is a can that holds 330 ml.

The cans are sold in packs holding 6, 10, 12 or 24 cans.

These packs are then stacked onto a pallet for delivery to the supermarket.

Larger bottles with capacities of 1.5 litres and 2 litres are usually sold as single units or in promotional packs which can vary in size from 2 to 8 bottles.

Visit a local supermarket and look at the soft drinks aisle.

Make a note of the different sizes and packages.

You will find different labels on the shelves telling you the cost of each product.

Is it easy to compare prices?

If not, why not?

Is bigger always better value?

If there is not a supermarket near you, you could use the internet to source your information.

Write a short report suggesting which product to recommend to a customer based on cost, capacity and portability of the product.

⟨GO!⟩ Activity

The table below shows the top selling DVDs in the USA in 2013.

Position	Title of film	Number sold
1	*The Twilight Saga: Breaking Dawn Part 2*	4 754 505
2	*Wreck-It Ralph*	2 814 660
3	*Hotel Transylvania*	2 655 365
4	*Taken 2*	2 615 088
5	*The Hobbit: An Unexpected Journey*	2 425 568
6	*Skyfall*	2 341 689
7	*Pitch Perfect*	2 209 207
8	*Rise of the Guardians*	2 037 123
9	*Lincoln*	1 755 708
10	*Les Misérables*	1 657 972

Source: http://www.the-numbers.com/home-market/dvd-sales/2013

The pitch at Hampden Park stadium measures 105 metres by 86 metres.

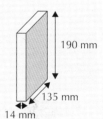

a If all of the number 1 selling DVDs were laid flat on the ground so as to cover the pitch, approximately how high would the DVDs reach?

b i If all the *Skyfall* DVDs were stacked on top of each other (cover side down), how high would the stack be?

   ii Find two cities which are this distance apart.

c If each film is, on average, 90 minutes long, how many hours would be spent in total watching **all** of the top 10 DVDs sold in 2013? Give your answer in years.

Using algorithms for efficient container packing

In Example 13.1, all the items to be packed were the same shape and size as each other.

In many situations, however, the items to be packed are different shapes and sizes. This makes the problem of fitting them into the container in the most efficient way more complicated.

Two different **algorithms** are commonly used to tackle such problems:

- first-fit algorithm
- first-fit decreasing algorithm.

In the **first-fit algorithm**, you take the items to be packed in the order they are given and fit each one into the first available container that will take it.

> ⚠ An algorithm is a step-by-step set of instructions that are followed until a particular outcome is achieved.

In the **first-fit decreasing algorithm**, you sort the items to be packed into order of decreasing size, then apply the first-fit algorithm.

Box I is too large for load 4, so will start load 5, leaving 110 cm.

Box L will also go into load 5, leaving 60 cm.

Box H will fit into load 4, filling it.

Box B will fit into load 5, leaving 30 cm.

Box M will fit into load 5, and will fill it.

The table shows the way Colin will pack the boxes for delivery using the first-fit decreasing algorithm.

Load	Boxes	Total depth (cm)
1	C + A	110 + 50 = 160
2	J + D	90 + 70 = 160
3	F + G	80 + 75 = 155
4	E + K + H	60 + 60 + 40 = 160
5	I + L + B + M	50 + 50 + 30 + 30 = 160

Using the first-fit decreasing algorithm, Colin can pack the boxes so as to make five trips instead of six. So this is the most efficient way for Colin to pack the boxes.

Note that in this arrangement, there is 'spare' space in only one load.

Exercise 13B

★ 1 A small ferry runs between an island and the mainland.

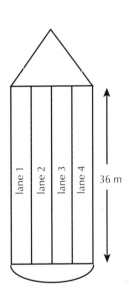

The ferry has four parking lanes to transport vehicles. Each lane is 36 m long.

The ferry company allows 9 m for each lorry, 5 m for each van and 4 m for each car booked onto the ferry.

The following bookings have been taken for the Monday afternoon ferry:

L, V, V, L, C, V, C, C, L, V, L, V, C, C, C, L, C, C, V, V, C, V, V, C, C, V

where L = lorry, V = van and C = car.

a Use the first-fit algorithm to try to find a suitable order in which to load the vehicles onto the ferry.

b Will the first-fit decreasing algorithm give a better arrangement? Give a reason for your answer.

2 For health and safety reasons, cartons must not be stacked to heights greater than 8 feet in a warehouse.

Thirteen cartons must be stacked in the most efficient way possible.

The heights of the cartons, in feet, are shown. All the cartons have the same sized square base.

4, 3, 7, 6, 5, 2, 3, 5, 8, 6, 2, 5, 4

a Use the first-fit algorithm to show a possible way to stack the cartons safely.

b Use the first-fit decreasing algorithm to find the most efficient way to stack the cartons.

⚙ 3   A small grocery shop has three shelves in its stockroom. Each shelf measures 30 cm by 180 cm and can take a maximum weight of 48 kg.

The owner of the shop goes to the cash and carry to buy stock for the shop. The table shows the products the shop owner bought, the quantities, the dimensions of their packaging and their weight.

Product	Number bought	Dimensions ($l \times b \times h$) cm	Weight (kg)
Tray of tinned tomatoes	6	$30 \times 30 \times 12$	5.2
3-litre cans of cooking oil	12	$12 \times 12 \times 22$	3.1
4-packs of orange juice	10	$11 \times 20 \times 22$	4.2
2 kg bags of sugar	14	$9 \times 12 \times 19$	2

What is the best possible way to arrange the goods on the shelves? Show your calculations.

⚙ 4   Jim has four vegetable beds in his garden.

Two of the beds measure 8 m by 3 m. The other two each measure 4 m by 4 m.

Jim wants to plant 30 m^2 of potatoes, 10 m^2 of cabbage, 6 m^2 of carrots, 2 m^2 of beetroot, 12 m^2 of broad beans, 8 m^2 of onions, 6 m^2 of courgettes and 6 m^2 of lettuce.

Apart from the potatoes, he does not want to plant any vegetable in more than one bed if possible.

Find a way for Jim to plant all the vegetables that he wants. Justify your solution by showing your calculations.

★ 5   The Dunfast Bunkhouse has eight rooms containing bunk beds.
⚙

Four of the rooms have eight bunk beds in them, two have six bunk beds in them and two have four bunk beds in them.

Each room can accommodate either men or women. There are no 'mixed' rooms.

The owner does not split up groups of tourists that book into the Bunkhouse, other than putting men and women in separate rooms.

Six groups are booked into the Bunkhouse one weekend. The table shows the number of women and men in each group.

Group	Women	Men
1	4	2
2	5	6
3	6	5
4	4	4
5	2	2
6	3	5

How could the owner accommodate the tourists in the rooms available? You must show your strategy and calculations.

- I can analyse a problem involving container packing.
  ★ Exercise 13A Q6   ⬭ ⬭ ⬯

- I can use appropriate strategies to find more than one solution. ★ Exercise 13A Q3, Q8   ⬭ ⬭ ⬯

- I can determine the most efficient solution based on the use of the first-fit and first-fit decreasing algorithms.
  ★ Exercise 13B Q1, Q5   ⬭ ⬭ ⬯

For further assessment opportunities, see the Case Studies for Unit 2 on pages 255–257.

14 Using precedence tables to plan tasks

This chapter will show you how to:

- use a precedence table to plan events where some activities can be done at the same time and others must be done in sequence
- use a network diagram to plan activities efficiently
- use critical path analysis to plan activities efficiently.

You should already know:

- why time management is important in everyday life
- how to create a schedule for personal activities
- how to plan for future events and justify the timetable created.

Putting tasks in order

Think about your daily routine and all the things you do. Every activity has an order which you follow without giving it any thought.

To complete a task it is sometimes necessary to complete one part before another can begin. We say that one task takes **precedence** over the other.

The tasks can be organised in various ways. These include **precedence tables**, **Gantt diagrams** and **network diagrams**. This chapter looks at each of these ways of organising tasks.

In each case, it should be possible to decide on the best order to undertake the tasks and the shortest time it will take to complete all the tasks, using the diagram that has been constructed. This is called the **critical path**.

Precedence tables

A precedence table can be used to plan events involving different numbers of activities.

In the table on the right:

- tasks A, B and C must be completed in sequence
- B can only be started when A is finished
- C can only be started when B is finished.

Task	Preceded by
A	–
B	A
C	A, B

In the table on the right:

- tasks A and C can be started at any time
- B can only be started when A is finished
- D can only be started when C is finished.

Task	Preceded by
A	–
B	A
C	–
D	C

Example 14.1

Goodfit Plumbers claim that they can refit a bathroom in 2 days.

The Goodfit team, which includes all the tradesmen needed to complete the job, works for 8 hours each day.

The table shows the tasks they need to complete and how long each will take.

Task	Activity	Preceded by	Time (hours)
A	Remove old bathroom fittings		3
B	Remove old floor tiles		2
C	Install new fittings	A, B	6
D	Lay new floor	A, B, C	4
E	Paint woodwork	A, B, C, D	2
F	Install window blinds	A, B, C, D, E, G	1
G	Paint walls	A, B, C, D	3

Draw a network diagram and work out the critical path to see if the bathroom be refitted in 2 working days as Goodfit Plumbers claim.

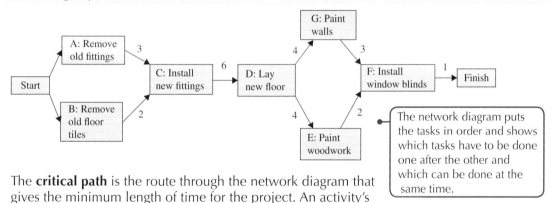

The **critical path** is the route through the network diagram that gives the minimum length of time for the project. An activity's **float** time is the amount of time an activity can be delayed without it affecting the overall time of the project. The float time of a critical activity is zero.

In this case, the critical path is:

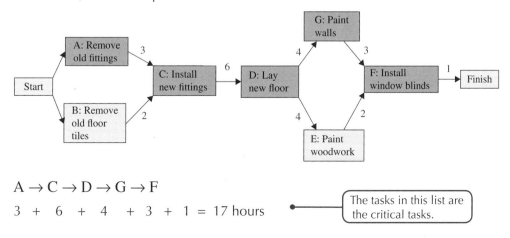

$A \rightarrow C \rightarrow D \rightarrow G \rightarrow F$

$3 + 6 + 4 + 3 + 1 = 17$ hours

The tasks in this list are the critical tasks.

The other tasks, B and E, are sub-critical and have a float time of 1 hour each.

This means that they do not need to be started as soon as the preceding task has been completed but could be postponed for up to 1 hour each.

The critical path shows that it will take 17 hours to complete the refit of the bathroom, so Goodfit Plumbers' claim made is not correct.

Exercise 14A

1   Kirsten is 17 and wants to learn to drive.

   She makes a list of the tasks that she will need to complete to get her full driving licence.

   She must hold a provisional driving licence before she can sit her theory test.

   She must pass the theory test before she is allowed to sit the practical test.

   She cannot apply for her full driving licence until she has passed the practical test. This is Kirsten's list.

   - *Apply for a full driving licence*
   - *Find a driving instructor*
   - *Take 10 weekly driving lessons*
   - *Book practical test*
   - *Pass theory test*
   - *Apply for a provisional driving licence*
   - *Hold a provisional driving licence*
   - *Learn theory*
   - *Book theory test*
   - *Pass practical test*

   a   Draw a precedence table for Kirsten with these headings:

Task	Description	Preceding task

   b   Put the tasks in order and complete the Preceding task column.

   c   How many different tasks have to be completed to get a full driving licence?

   d   Can any of the tasks be done at the same time? If so, which?

2 Imagine it is Monday morning. Your alarm clock has just gone off and you have to get yourself organised and ready for school. Some of the tasks you are likely to do are listed below.

get dressed   make packed lunch   clean shoes

eat breakfast   check phone   find PE kit

have a shower   pack school bag   find school uniform

    a Put these activities in a logical order by considering if some things need to be done before others.

    b Write down any other things that you do in the morning which are not included in this list.

    c List any of these activities which could be completed the night before.

    d Write down why it would be useful to complete these activities earlier.

    e Is there another way in which you could get these things completed more quickly?

3 Devindra is a roofer. He makes this list of the things that he and his partner do when they go to do a job:

Load the van

Send an invoice

Order scaffolding

Order tiles and fixings

Check the weather

Arrange time to be on site

Initial visit to size up the job

Mail out quote

Fix any snags

    a Copy and complete the table below by putting the activities into a logical order.

        Identify when two or more activities could be done at the same time.

        Put the second activity in the right-hand column.

Order of activities	Activities that could be done at the same time

    b Draw a network diagram to show the tasks in order.

★ 4 Yasmeen is making lunch of soup, a sandwich and a glass of orange juice. This involves the following tasks (times are given in minutes).

- Opening a tin of soup (3)
- Putting the soup in a saucepan (2)
- Heating the soup (5)
- Slicing the bread (3)
- Buttering the bread (3)
- Filling the sandwich with ham and salad (4)
- Getting plates and a glass out of the cupboard (3)
- Pouring juice into the glass (1)
- Serving lunch (3)

Task	Description	Preceding task	Time
A			
B			
C			
D			
E			
F			
G			
H			
I			

a Copy the table above and put the tasks in order, showing which tasks are dependent on others.

b Use the table to create a network diagram showing which tasks can be done at the same time.

c Use the critical path to find the least amount of time it will take Yasmeen to prepare lunch.

5 Rebecca is going to redecorate her bedroom and buy some new furniture for it. The table lists the tasks she must carry out.

Task	Description	Preceding task	Time (hours)
A	Clear old furniture and carpet from room	–	2
B	Paint the woodwork	A	2
C	Hang the wallpaper	A, B	6
D	Lay the new carpet	A, B, C	2
E	Assemble the new furniture	–	4
F	Move furniture into the bedroom	A, B, C, D, E	1

a Which tasks could be done at the same time?

b Rebecca plans to do all the work herself. Draw a network diagram to show the order in which the tasks should be completed and to show which tasks could be done at the same time.

c Use the critical path to find the minimum amount of time needed to complete the bedroom. You must explain your answer.

6 Andrew and Marina are doing some housework. They have agreed to complete the jobs described below.

Task	Description	Preceding task	Time (minutes)
A	Load the dishwasher	–	8
B	Run dishwasher cycle	A	70
C	Sort dirty clothes in lights and darks	–	5
D	Put light clothes in washing machine	C	3
E	Run washing machine cycle	C, D	85
F	Empty the dishwasher and put away dishes	A, B	15
G	Empty the washing machine	C, D, E	3
H	Hang up clothes to dry	C, D, E, G	10

Write down the two household tasks they are doing.

a Andrew and Marina plan to go out when both the jobs are completed.

They start the jobs at 0930. What is the earliest time that they can go out?

b If Marina has to do all the tasks herself, how long will it take her? You must explain your answer.

7 Tony is going to tackle the pile of clothes lying in the corner of his bedroom. He will keep the clothes he still likes, get them washed, ironed and put away neatly in his wardrobe. He will put any unwanted clothes in a bag to take to the local charity shop.

a Decide on the tasks Tony needs to do to complete this job.

b Draw and complete a precedence table to illustrate this with an estimated time for each task.

c Describe how Tony could complete this job in the shortest time.

8 Lauren has invited friends over for a meal at 8 pm on Saturday night but hasn't yet planned what to cook. She needs to get organised.

a Think about your favourite meals and decide on a main course and dessert to recommend to Lauren.

b List all the tasks required to prepare and make each course and complete a precedence table with times to illustrate this.

c Describe how Lauren could prepare her meal in the quickest time.

d Lauren's partner says that he will make the main course. Does this change the total time required to prepare the meal?

e What is the latest time that she can start preparing the meal to ensure that the main course is ready to serve at 8.15 pm?

★ 9 At 4.30 pm the resignation of a government minister is announced. A TV network wants to
prepare an item about the story for the 6 o'clock news. The following information describes the
tasks involved in preparing the news item.

Interview the resigning minister 15 minutes	Get reaction from former colleagues 25 minutes	The resigning minister will be interviewed at 4.30pm
Review the minister's career 25 minutes	Prepare film from archives 20 minutes	Edit all the materials 20 minutes
Final task: edit the film when all other tasks are complete	You cannot prepare the film from archives until you have reviewed the minister's career and spoken to his colleagues	You must interview the minister before reviewing his career.
Review possible replacements for the minister 40 minutes	You must review possible replacements after filming Downing Street.	Film Downing Street 20 minutes

You cannot get the reaction from colleagues until you have interviewed the minister and
completed the filming at Downing Street

a Use the information to plan and order all the tasks required to ensure that the news item is
ready for broadcast on the 6 pm news.

b Illustrate this information in a precedence table.

c Each of the jobs before the final edit needs a reporter. Once a reporter has started a job they
must complete it. Explain why a minimum of three reporters are needed to have the news
item ready before 6 pm.

10 Sandra employs a firm called Waterises to fit a new bathroom. The firm uses a team of workers to complete the installation. The table shows the tasks and the time needed for each.

Task	Detail	Preceding task	Time (hours)
A	Remove old suite	–	3
B	Build cupboards	–	1
C	Insert new plumbing	A	6
D	Remove old tiles	A	1
E	Begin electrics	A, D	2
F	Plaster walls	A, D, E,	7
G	Fit new suite	A, C, D, E, F	8
H	Fit cupboards	A, B, D, E, F	1
I	Tile walls	A, B, C, D, E, F, G, H	6
J	Finish electrics	A, B, C, D, E, F, G, H, I	1
K	Finish plumbing	A, B, C, D, E, F, G, H, I	1
L	Finish decoration	A, B, C, D, E, F, G, H, I	5

a Create a network diagram to show the tasks in the table.

b Waterises claim they can complete the job in 30 hours. Do you agree with this claim? Give a reason for your answer.

GO! Activity

Research the following online or by speaking to friends and family.

* What is the current cost of a provisional driving licence?

* At what age can you apply for this?

* How do you find a driving instructor?

* What is the current cost of a typical, single driving lesson?

* How do you learn the theory required for the theory test?

* How many questions are in the theory test? What kind of response is needed?

* How do you take the theory test? What is the pass mark?

* On average, how many lessons do you need before you are ready to take your practical test?

* On average, estimate how much it will cost to learn to drive. Justify your answer with appropriate calculations.

GO! Activity

1. Your school is putting on a musical. This involves many jobs, each of which take different times to be completed. These jobs may include:

- Make the costumes
- Conduct rehearsals
- Get posters and tickets printed
- Run competition to design a poster
- Get programme printed
- Make scenery and props
- Choose cast
- Get rights to perform the musical
- Hire hall
- Arrange refreshments for show
- Organise make up

- Decide on the show
- Organise lighting
- Conduct dress rehearsals
- Invite local press and radio
- Choose stage hands
- Choose poster design
- Choose programme sellers
- Decide on date of show
- Arrange seating
- Sell tickets
- Display posters

   a Illustrate this information in a precedence table. Complete the precedence relationships and estimate the time you think it will take to complete each job.

   b By considering your precedence table and estimated times, what is the minimum time required to be ready for opening night?

- I can plan an event by putting tasks in logical order, working out which can be done at the same time, and so work out the minimum time needed for the event.
  ★ Exercise 14A Q4

For further assessment opportunities, see the Case Studies for Unit 2 on pages 255–257.

15 Solving a problem involving time management

Planning your time

Time management is a very important mathematical skill. As you get older and become more independent, it will become increasingly important that you can organise your day so that you can include all the tasks that you need to complete.

School probably takes up most of your time each day, but you still need to be able to plan your evenings and weekends to fit in hobbies, sport, meeting your friends and spending time with your family.

Planning ahead will ensure that you can do all the things that you want to do and that you don't take on too many commitments that you will be unable to fulfil.

An important skill is being able to estimate how long it will take to complete a job or a piece of work.

If you overestimate, you will finish the job early and you could end up wasting time.

If you underestimate, you will take longer than planned to complete the job and other things that you had planned to do will be delayed or possibly cancelled.

If you are running a business either of these could cause you to lose money. You could have staff with nothing to do or you could have to pay staff overtime to complete jobs that take longer than planned.

Example 15.1

Ross runs a gardening company. He has been asked to quote how much it will cost for him to cut an 'average' grass lawn measuring 10 m by 8 m.

Ross charges £20 per hour. He charges a minimum of 1 hour. He then charges for each additional 15 minutes worked. Time worked over an hour is rounded up to the nearest 15 minute interval.

a How much should he charge for mowing an 'average' lawn?

b How might his quote vary for a lawn that is not 'average'?

c Why is it important that Ross try to estimate the job accurately?

a Ross works out an 'average' time for each part of the job:

- take edging shears and lawnmower out of van and top up fuel – 10 minutes
- cut the grass – 30 minutes
- empty grass cuttings onto compost heap – 10 minutes
- cut the edges manually with edging shears – 15 minutes
- put lawnmower and edging shears back into van – 5 minutes

> By breaking down the task into its individual parts he can get a better estimate of the time needed for the whole activity.

Total estimated time is 1 hour and 10 minutes.

Charge: £20 + £5 = £25

> The charges for 1 hour and for 15 minutes.

b For a larger garden or for a lawn that is an awkward shape, Ross should increase the time for actually cutting the grass and, perhaps, for emptying the grass cuttings into the compost heap. The other times will stay the same.

c If his time estimate is too low, then he will charge too little and not earn enough.

If his time estimate is too high, then he will overcharge the customer, who might then decide that he is too expensive and next time decide to employ another gardener.

Exercise 15A

1 **a** Draw a timeline for all the activities that you did during a 24-hour period from 9 am on a weekday this week.

b Which activity took up most of your time?

c Which activity took up the least of your time?

d Draw a suitable graph to illustrate your results.

e Would your timeline be the same at the weekend as on a school day? Draw a second timeline to show 24 hours on a typical weekend and draw a graph to illustrate these results.

2 Estimate how long it would take you to complete the following tasks:

a meet friends to watch a film at the cinema

b do the ironing

c prepare and eat a three course meal.

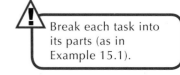

> Break each task into its parts (as in Example 15.1).

3 Morag and Ali live in Aviemore and are planning to go Christmas shopping together in Edinburgh. They have both created a schedule for the day.

Morag's plan	
0800	Start drive to Edinburgh (130 miles)
1045	Arrive at Hermiston Park and Ride, Edinburgh (free parking)
1055	Catch bus to Princes Street
1135	Arrive Princes Street
1135	Shop
1300	Lunch at Pizza Shack
1400	Shop
1600	Catch bus back to Hermiston Park and Ride
1650	Leave Park and Ride to drive home (130 miles)
1930	Arrive home

Ali's plan	
0830	Catch train from Aviemore to Edinburgh
1117	Arrive Edinburgh Waverley (on Princes Street)
1130	Coffee
1200	Shop
1300	Lunch at Pizza Shack
1400	Shop
1545	Coffee
1633	Catch train from Edinburgh Waverley to Aviemore
1928	Arrive in Aviemore

Morag's car costs 20p per mile to run.

Bus fares in Edinburgh cost £3.50 per person for an all day ticket.

The train fare from Aviemore to Edinburgh is £16.50 per person.

The train fare from Edinburgh to Aviemore is £13.50 per person.

a What are advantages and disadvantages of each plan?

b How much shopping time do you get with each plan?

c Which plan do you prefer and why?

4 a Make a list of all the things you have to do over the next 7 days in and out of school.

b Prioritise your 'to do' list from most important to least important.

c Put a star next to the things you are looking forward to doing.

d Put a circle next to the things you have to do, but are not so keen on.

e Make up a schedule for the next 7 days to plan how you will tackle your 'to do' list. Try to estimate how long each event will take to allow enough time to complete them. Mix up tasks with stars and circles, to avoid leaving all the circles until the end of the week.

A school planner or diary will be useful for this question.

Planning events

As an adult, it will be important that you can plan for big events like a holiday or a wedding.

Big events may involve you individually or may involve a group of family, friends or colleagues. In such situations you will have many factors to consider including:

- the individual tasks needed to prepare for the event

- how long each task will take

- the total time to be planned for

- contingencies in case something goes wrong.

It is often a good idea to create a timeline for the event that you are planning.

As discussed in Chapter 14, it might be possible to shorten the total time needed for the task by completing some of the subtasks at the same time.

Example 15.2

A builder is constructing a house and has listed the different jobs involved, along with estimated durations:

A   Do outer brickwork – 5 days

B   Landscape site – 2 days

C   Order windows and other materials and wait for delivery – 2 weeks

D   Remove scaffolding – 1 day

E   Dig foundations, pour concrete and wait to go off – 3 weeks

F   Slate roof – 7 days

G   Fit windows – 1 day

H   Prepare site and install scaffolding – 4 days

I   Mount roof trusses – 2 days

a   Put the jobs in order and work out the length of the build.

b   Identify any jobs that can be done at the same time and work out the new estimated time for the build.

a   The jobs can be completed in the following order:

E, C, H, A, I, F, G, D, B

This would give a total time of 57 days.

b   Tasks E, C and H can be done at the same time, as can F and G. This can be shown in a diagram:

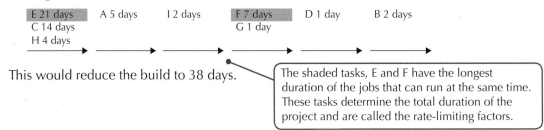

E 21 days   A 5 days   I 2 days   F 7 days   D 1 day   B 2 days
C 14 days   G 1 day
H 4 days

This would reduce the build to 38 days.

> The shaded tasks, E and F have the longest duration of the jobs that can run at the same time. These tasks determine the total duration of the project and are called the rate-limiting factors.

Exercise 15B

1   Look at the tasks for making beans on toast.

A Put beans in a pan and heat	B Put toast onto a plate	C Put bread in toaster	D Put beans on toast and serve
E Get plate and cutlery ready	F Open tin of beans	G Get tin of beans from cupboard	H Take bread out of freezer

a   Put the tasks in order.

b   Which tasks can be done simultaneously?

c   Estimate how long each task takes.

d   How long should it take to make the beans on toast?

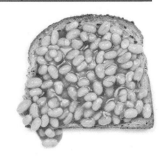

★ 2 David and Helen have bought a house that needs to be completely renovated. They decide to employ a team of builders to do all the work. Helen makes a list of tasks to be completed.

A. *Install kitchen units – 3 days*	**B.** *Move in furniture – 1 day*
C. *Lay laminate flooring – 2 days*	**D.** *Remove old kitchen – 1 day*
E. *Plaster walls – 3 days*	**F.** *Lay carpets – 1 day*
G. *Remove old bathroom – 2 days*	**H.** *Tile kitchen and bathroom walls – 3 days*
I. *Rewire the house – 5 days*	**J.** *Install new bathroom – 3 days*
K. *Remove old carpets and sand floors in living room – 3 days*	**L.** *Paint walls and ceiling – 3 days*
M. *Hang curtains and blinds – 1 day*	**N.** *Install plumbing – 4 days*

a Working with a partner, put the jobs in order. Discuss and write down any jobs that can be done at the same time, making a note of your reasons.

b Estimate the total time to complete the project.

3 Yasmine, Tia and Duncan live in Pitlochry. They are planning to visit a theme park in Motherwell, 83 miles away, during the Easter holidays.

They plan to drive there and back in one day.

They want to arrive in time for the park opening at 0930.

The theme park has shows as well as rides.

They want to see all the shows listed below.

They plan to leave as soon as the last show finishes so that they are not too late getting home.

Show	Start times	Length of show (min)
Star Maker	1215, 1400, 1545	45
Penguin Parade	1245, 1430	30
Show Time	1115, 1315, 1515, 1715	35
Out of This World	1245, 1545	40

a Yasmine thinks that they will be able to travel at an average speed of 40 mph. What time should they leave home so that they arrive at the theme park by 0930?

b Due to heavy traffic, their actual average speed on the way to the theme park is 25 mph. What time will they actually arrive at the theme park (use the start time in part a to calculate your answer)?

c What will be the best times for them to see each of the four shows?

d Unfortunately, it takes 40 minutes for the group to return to their car and leave the car park as it is busy. They manage an average speed of 46 mph on the way home. Will they be home in time to see the final episode of their favourite comedy series which starts at 1930?

GO! Activity

You are planning a special meal for Father's day. You are going to cook a three-course meal.

What will you cook and how long will it take to prepare and cook the entire meal?

a Decide on what you will cook for each course.

b Make a list of all the tasks needed to make all three courses of your meal.

c Estimate the total time for preparation and cooking.

d You plan to eat at 7 pm. Use your answers to **b** and **c** to make an approximate timeline so that everything is ready on time.

World time zones and 24-hour time

There are 24 *time zones* around the world. In each time zone, noon (1200) is approximately the time when the sun is highest in the sky.

All of the world's time is set from Greenwich in London and this central time zone is called Greenwich Mean Time (GMT).

Countries to the east of London are ahead of GMT while countries to the west of London are behind GMT.

When you arrive in a foreign country you often have to re-set your watch so that it shows the correct **local time** for that country.

The table below shows the time difference (ahead of or behind GMT) for a number of major cities around the world.

Country	+ GMT or – GMT	Country	+ GMT or – GMT
Las Vegas, USA	– 8 hours	Moscow, Russia	+ 3 hours
Denver, USA	– 7 hours	Lahore, Pakistan	+ 5 hours
Chicago, USA	– 6 hours	New Delhi, India	+ 5 hours 30 min
New York, USA	– 5 hours	Beijing, China	+ 8 hours
Mexico City, Mexico	– 2 hours	Hong Kong	+ 8 hours
Brasilia, Brazil	– 2 hours	Tokyo, Japan	+ 9 hours
Paris, France	+ 1 hour	Brisbane, Australia	+ 10 hours
Madrid, Spain	+ 1 hour	Sydney, Australia	+ 10 hours
Athens, Greece	+ 2 hours	Wellington, New Zealand	+ 12 hours

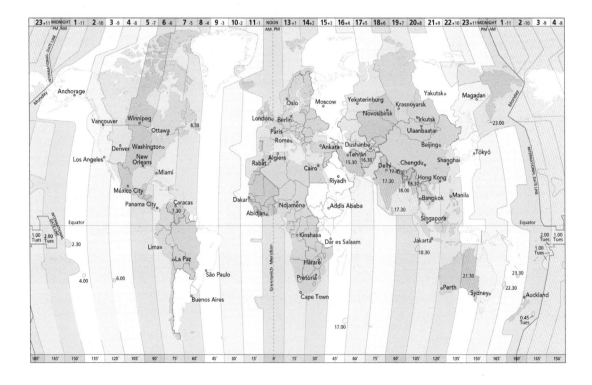

Example 15.3

Michael goes to Australia to visit his brother in Sydney.

His plane leaves London at 1430 on Tuesday 13 January.

First he flies to Hong Kong for a stopover and then he flies from Hong Kong to Sydney.

The flight from London to Hong Kong takes 11 hours 42 minutes.

His stopover in Hong Kong lasts for 3 hours 20 minutes.

The flight from Hong Kong to Sydney takes 9 hours 38 minutes.

a   At what local time does Michael's plane arrive in Hong Kong?

b   At what local time does his plane leave Hong Kong for Sydney?

c   At what local time does Michael arrive in Sydney?

a   1430 (Tuesday) + 11 h 42 min = 0212 (Wednesday) GMT

Michael's plane lands at
0212 + 8 h = 1012 (local time) Wednesday 14
January in Hong Kong.

> Add the flight time to the departure time to work out the equivalent GMT when the flight lands in Hong Kong.

> From the table, Hong Kong is 8h ahead of GMT, so add 8h to GMT.

b   Michael's plane leaves Hong Kong
at 1012 + 3 h 20 min =
1332 local time.

> Add the stopover time of 3h 20min to the arrival time.

c   1332 (Wednesday) + 9 h 38 min = 2310
(Wednesday) Hong Kong time

Michael's plane lands in Sydney at
2310 + 2 h = 0110 Thursday
15 January in Sydney.

> From the table, Sydney is 10h ahead of GMT, which makes it 2h ahead of Hong Kong time. Add 2h to the equivalent time in Hong Kong.

Exercise15C

1   In Edinburgh the time is 1300. What time is it in these cities?

   a   Athens   b   Moscow

   c   Hong Kong   d   Mexico City

2   The 2014 Winter Olympics took place in Sochi, Russia. Sochi is in the same time zone as Moscow. The opening ceremony was scheduled to start at 2000 local time and was shown live around the world. What was the local time in these cities when the opening ceremony started?

   a   Inverness   b   Lahore   c   Las Vegas

3   The final of the 20th FIFA World Cup took place in Brazil at 4 pm local time on 13 July 2014. The match was shown live around the world. What was the local time in these cities when the final match started?

   a   New York   b   London

   c   Tokyo   d   Sydney

★ 4   a   If it is 2230 on 12 March in New York, what time and date is it in Moscow?

     b   If it is 3 am on 29 December in Brisbane, what time and date is it in Chicago?

5 **a** A flight from Edinburgh to Athens takes 6 hours 45 minutes. If the plane takes off at 1.30 pm, what will be the local time when it lands?

b A flight from London to New Delhi takes 12 hours. If the plane takes off at 1730, what will be the local time when it lands?

6 A flight from Edinburgh to Sydney, Australia, via Hong Kong takes 24 hours in total.

a If the plane takes off at 3 pm on 15 December, what is the date and local time when it lands in Sydney?

b The flight distance is approximately 10 500 miles. Calculate an estimate of the average speed for the journey.

★ **7** A plane leaves London at 0915 on 7 January to fly to Wellington, New Zealand. The journey, including two stops, takes 42 hours 25 minutes.

a What is the date and local time when the plane lands in Wellington?

b The flight distance is approximately 11 600 miles. The total time spent in airports during the stopovers was 8 hours. Calculate an estimate of the average flying speed of the plane for the journey.

8 Calculate the actual flight time for each of these trips.

	Departure city	Departure time (local time)	Arrival city	+ GMT or – GMT	Arrival time (local time)
a	London, UK	1120 Saturday	Vancouver	– 8 hours	1240 Saturday
b	London, UK	1000 Saturday	Toronto	– 5 hours	1215 Saturday
c	London, UK	1040 Saturday	Cape Town	+ 2 hours	2040 Saturday
d	London, UK	1125 Saturday	Dubai	+ 4 hours	2135 Saturday

9 Alana flies from London to New York for a business meeting. The company arranges a car to pick her up at the airport in New York.

The flight to New York takes 8 hours. The flight leaves London at 1500. Alana allows 2 hours 30 minutes to get through immigration and to pick up her luggage.

What time should the car be at the airport in New York to pick up Alana?

★ **10** A company has offices in London, Paris and New York. Each office is open from 0800 until 1700 local time. The managers of the three offices need to arrange a video conference at a time when all three offices are open. The conference will last for 1 hour.

Between which times can the video conference be held? Give the local times for each city.

11 Rohan is a self-employed courier. He has three packages to deliver on one day.

He needs to return home after his final delivery before leaving for a 2 pm appointment in Edinburgh. Each delivery takes 10 minutes to complete. He needs 20 minutes at home before leaving for Edinburgh. He wants to arrive 15 minutes before his appointment time.

The details of the distances he must drive and the average speed that he can travel for each part of his day are shown in the table.

Journey	Distance (miles)	Average speed (mph)
Home to delivery A	12	30
Delivery A to delivery B	20	40
Delivery B to delivery C	15	25
Delivery C to home	16	30
Home to Edinburgh	30	40

What is the latest time at which Rohan can leave home to start his deliveries if he is to be in time for his appointment in Edinburgh?

12 A company exports its products by cargo plane to Moscow in Russia, Bucharest in Romania, and Algiers in Algeria.

The plane must be on the ground for 3 hours at each airport to unload cargo and refuel.

The plane carries two pilots so that they do not need to have an overnight stay in any of the countries they are delivering to.

The cargo plane flies from Edinburgh and follows the course described in the table.

Departs	Arrives	Distance (km)	Bearing	Flying time	Time difference between point of departure and arrival
Edinburgh	Moscow	2500	090°	3 h 16 min	+3 hours
Moscow	Bucharest	1500	210°	1 h 58 min	− 2 hours
Bucharest	Algiers	2100	250°	2 h 45 min	− 1 hour

a Choose a suitable scale and create a scale drawing of the course taken by the plane.

b Use your scale drawing to find the distance and bearing of the return flight from Algiers to Edinburgh.

> ⚠ Look at Chapter 12 for information on bearings and scale drawings.

c The plane leaves Edinburgh to fly to Moscow on Thursday at 11.45 pm. Calculate the local landing and take-off times at each of Moscow, Bucharest and Algiers.

d The distance from Algiers to Edinburgh is approximately 2200 km. What is the date and local time when the plane lands back in Edinburgh? Use your calculations to justify your answer.

★ **13** Ashleigh flew from London to New Zealand to work on a farm near Auckland. These are the details of her flight:

- distance from London to Auckland: 18 330 km

- aircraft speed: 975 km/h

- duration of stopover in Hong Kong: 2 h 40 min

- time difference between London and Auckland: + 12 hours.

  a Calculate how long the journey from London to Auckland took, including the stopover in Hong Kong. Give your answer in hours and minutes.

  b Ashleigh left London at 1345 on Sunday 26 October 2014. What was the date and local time when she arrived in Auckland?

  c Ashleigh's father lives in Glasgow. He planned to phone Ashleigh at 1930 UK time on Monday 27 October. Was this a suitable time to contact Ashleigh? Explain your answer with appropriate working.

  d Ashleigh works from 0830 to 1800 Monday to Saturday, but doesn't work on Wednesday afternoons after 1 pm. She likes to be in bed by 2230. When are the best UK times for her father to phone her?

Activity

You have 6 weeks to travel around the world during your gap year before college. Think about the following:

- which countries you would like to visit

- how long you would stay in each country

- where you would stay in each country

- what key activities you would like to do in each country

- which points of interest you would like to visit in each country.

a Put together an itinerary detailing your travel plans and costs. Use the internet to find out flight times and accommodation details which meet your needs.

b Every Sunday one of your parents will ring you. Work out the most suitable UK time and local time (depending on where you will be) for them to call.

- I can plan activities and events making best use of the available time. ★ Exercise 15A Q4   ★ Exercise 15B Q2

- I can use time intervals to work out times in different time zones. ★ Exercise 15C Q4, Q7

- I can solve a time management problem which includes working across time zones. ★ Exercise 15C Q10, Q13

For further assessment opportunities, see the Case Studies for Unit 2 on pages 255–257.

16 Considering the effects of tolerance

This chapter will show you how to:

- calculate the limits of measurements given different levels of tolerance
- explain the use of appropriate levels of tolerance for fitting components
- solve problems involving tolerance calculations.

You should already know:

- how to round numbers to the nearest significant figure or a number of decimal places
- that different degrees of accuracy are acceptable in different situations
- how to interpret and write measurements given in tolerance notation.

Tolerance

When manufacturing a component it is impossible to achieve 100% accuracy in the production process. Instead the manufacturer will allow for a certain variation in accuracy.

By considering the size of the parts that the component must fit into, the manufacturer may be willing to accept a variation of, for example, 1 mm on either side of a 15 mm length.

We say that 15 mm is the **nominal** length.

The manufacturer will state the **tolerance** for the length of the component as ± 1 mm. The length of the component is written as (15 ± 1) mm. The minimum length accepted will be 15 − 1 = 14 mm and the maximum length will be 15 + 1 = 16 mm.

Example 16.1

The length of a new pencil has a tolerance of (18.0 ± 0.3) cm.

State the minimum and maximum acceptable lengths.

Minimum acceptable length is (18.0 − 0.3) cm = 17.7 cm

Maximum acceptable length is (18.0 + 0.3) cm = 18.3 cm

Example 16.2

A bolt is designed to fit into a given hole. The diameter of the hole has a tolerance range of 6.94 mm to 7.16 mm.

Write the diameter of the hole in tolerance notation.

7.16 − 6.94 = 0.22 mm — Work out the difference between the maximum and minimum diameters.

0.22 mm ÷ 2 = 0.11 mm — The tolerance is half the difference between the maximum and minimum diameters.

6.94 + 0.11 mm = 7.05 mm — The nominal diameter of the hole is the minimum acceptable diameter plus the tolerance.

Diameter is (7.05 ± 0.11) mm. — Write your answer using tolerance notation.

If tolerance is not given, it can be calculated by **having the lowest whole value of the stated unit of measurement**.

Example 16.3

Mr Cameron is having a new kitchen fitted in his house.

He plans to have five standard kitchen units fitted along one wall.

The nominal width of a standard kitchen unit is 60 cm.

What are the minimum and maximum possible total widths for the five units?

The width of each kitchen unit can be written as (60 ± 0.5) cm.

The minimum possible width of a unit is 59.5 cm and the maximum possible width is 60.5 cm.

> The least unit of measurement is 1 cm, so the tolerance will be half of this, i.e. 0.5 cm.

The minimum possible total width of five kitchen units is $5 \times 59.5 = 297.5$ cm.

The maximum possible total width of five kitchen units is $5 \times 60.5 = 302.5$ cm.

For very small objects such as micro-organisms or fibres, the standard units of measurement are too large. The **micron** or micrometre is often used instead. A micron is equal to one millionth of 1 metre. The symbol for micron is µm.

1 metre = 1 000 000 microns	1 mm = 1000 microns
1 m = 1 000 000 µm	1 mm = 1000 µm
1 µm = 0.000 0001 m	1 µm = 0.001 mm

Example 16.4

A strand of human hair is measured to have a diameter of 100 µm.

What are the minimum and maximum possible diameters for this strand of hair?

The minimum possible diameter is $(100 - 0.5)$ µm = 99.5 µm.

The maximum possible diameter is $(100 + 0.5)$ µm = 100.5 µm.

> The least unit of measurement is 1 µm, so the tolerance will be half of this, i.e. 0.5 µm.

Exercise 16A

★ 1 For each of the following measurements, write down the minimum and maximum acceptable sizes.

    a (250 ± 15) kg   b (152 ± 34) mm

    c (13.1 ± 0.3) mg   d (37.5 ± 0.5) ml

    e (8.25 ± 0.05) mg   f (0.653 ± 0.003) mm

    g (6.549 ± 0.005) mm   h (5.38 ± 0.03) µm

    i (0.74 ± 0.001) µm

2   Write each of the following using tolerance notation.

   a   minimum = 2.9 cm   maximum = 3.0 cm

   b   minimum = 97.6 g   maximum = 100 g

   c   minimum = 97.3 °F   maximum = 99.1 °F

   d   minimum = 1.35 kg   maximum = 1.45 kg

   e   minimum = 49.59 m*l*   maximum = 49.89 m*l*

   f   minimum = 1.234 µm   maximum = 1.258 µm

3   The diameter of a computer cable is given as (16 mm ± 10%).

   What are the minimum and maximum possible diameters for the cable?

4   The table shows the measured diameters of different types of fibre.

   a   Write each of these diameters in tolerance notation using microns.

   b   Write each of these diameters in tolerance notation using millimetres.

   c   How many times thicker is a human hair than an Angora rabbit fibre?

Fibre	Diameter (µm)
Rabbit angora	10–12
Cashmere	15–19
Alpaca	12–32
Mohair	23–45
Human hair	60–70

5   A baby's weight is given as 3.8 kg ± 0.05 kg. What are the minimum and maximum possible weights of the baby?

6   Lewis makes a tower using 10 bricks. Each brick has a height of (8.8 ± 3.0) mm. What are the minimum and maximum possible heights of the tower?

7   Thirty airline passengers check in one bag each at the airport. The bags are weighed separately. The scales at the airport are accurate to ± 500 g. The nominal total weight of the 30 bags is 580 kg. What is the maximum possible total weight of the 30 bags?

8   A section of gutter for a house has a nominal length of 4.00 m. Find the minimum and maximum possible total length of six sections of gutter.

9   Malcolm has a contract to fit television cables in 10 new houses. He buys (150 ± 0.25) m of cable. Each house needs (15 ± 0.5) m of cable. Does Malcolm have enough cable to fit television cables in all 10 houses? Show your calculations to support your answer.

GO! Activity

Excess baggage charges imposed on passengers at major airports can be due to faulty weighing scales. This could mean that large numbers of travellers have been unfairly charged excess fees. The most common fault identified at airports is that scales are set above zero before baggage is weighed.

a   By researching online, using a travel agent or travel brochures:

   •   investigate the maximum luggage allowance for each passenger on different airlines

   •   investigate the different charges of each airline for excess baggage.

b   If the scales at the airport have a tolerance of 500 g, what is the biggest weight that can be packed without being charged extra by each airline?

c   Write a short summary of your findings, comparing the restrictions that different airlines have.

Do tolerance levels always work?

It is very important to modern manufacturing, particularly with very small items, that components are produced to appropriate tolerances.

Example 16.5

A company building a new Formula 1 car decides to order all the nuts and bolts with dimensions (5.0 ± 0.1) mm.

What problems could this cause?

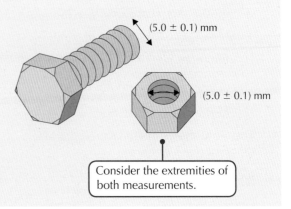

(5.0 ± 0.1) mm

(5.0 ± 0.1) mm

Consider the extremities of both measurements.

The bolts could have a diameter from 4.9 mm up to 5.1 mm.

The nuts would also have a diameter from 4.9 mm to 5.1 mm.

If a bolt has been supplied at 5.1 mm and a nut at 4.9 mm, the bolt would be too wide to fit the nut.

Similarly, if the bolt was 4.9 mm and the nut 5.1 mm, the bolt would be too small to fit the nut correctly.

It should be possible to find a nut to fit all the bolts, but the manufacturer would not want to waste the time and money that would be needed to search through the supplied parts looking for a 'good fit'.

Exercise 16B

1   Rohanna wants to order rigid plastic storage boxes to fit into her wardrobe. She has accurately measured the space available in her wardrobe to be 1.61 m wide.

   The storage boxes that she wants are available in different widths. The nominal widths are 30 cm, 40 cm and 50 cm each at ± 0.5 cm.

   Which boxes should she order to guarantee that they will fit the wardrobe using space most efficiently? Justify your solution.

★ 2   Ian is planning to replace the doors in his bedroom and kitchen. He chooses doors that have a width of (76 ± 1) cm.

   The bedroom door frame has a width of (75 ± 1) cm and the kitchen door frame has a width of (77 ± 1) cm.

   Will he be able to use the same style of door for both rooms? Use your working to explain your answer.

3   A steel beam measuring (4.6 ± 0.02) m long is fitted between two joists which are (4.7 ± 0.1) cm apart.

   Can the beam always be fitted between the joists? Use your working to explain your answer.

4 A soft drinks firm produces cans of cola which have a volume of $(330 \pm 10)\,\text{m}l$.

If more than 15% of the cans have a volume outside the stated tolerance, the production line is stopped so that the machines can be recalibrated.

A random sample of eight cans is taken and their volumes measured accurately. The volumes of these cans, in ml, are:

329, 341, 319, 326, 330, 328, 326, 335

Should the production line be closed down for recalibration? Use your working to explain your answer.

5 A saw mill produces planks of wood of length $(200 \pm 2.5)\,\text{cm}$.

The saw mill adjusts the machinery if more than 20% of the planks produced are not within tolerance.

A random sample of 10 planks is taken. The lengths of these planks, in cm, are:

203.0, 201.5, 198.2, 197.5, 197.3, 202.4, 199.2, 200.6, 202.3, 201.8

Does the saw mill need to adjust the machinery? Use your working to justify your answer.

Calculations involving tolerance

It is important to understand the cumulative effect that tolerances can have on calculations that involve several different measurements.

If all the individual measurements are at the upper end of the range, then the final result could be much larger than if all the measurements are at the lower end of the range.

This has implications for the accuracy of the final answer.

Example 16.6

a Mr McGregor measures his rectangular vegetable patch to the nearest metre. It measures 12 m by 7 m.

    i What are the minimum and maximum possible areas of the vegetable patch?

    ii Write the area of the vegetable patch in tolerance notation.

b Fertiliser is available for the garden in $(10 \pm 0.3)\,\text{kg}$ tubs. The garden needs 110 g of fertiliser per square metre. Will one tub of fertiliser be sufficient for the whole garden?

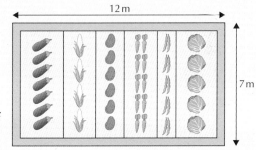

a **i** Minimum length is 11.5 m.

    Minimum breadth is 6.5 m. •————— *The least unit of measurement is 1 m, so the tolerance will be half of this, i.e. 0.5 m.*

    Minimum possible area is $11.5 \times 6.5 = 74.75\,\text{m}^2$.

    Maximum length is 12.5 m. •————— *The minimum area of the vegetable patch will be when the length and breadth are both at the minimum possible measurements.*

    Maximum breadth is 7.5 m.

    Maximum possible area is $12.5 \times 7.5 = 93.75\,\text{m}^2$.

(continued)

ii $93.75 - 74.75 = 19\,m^2$

$19\,m^2 \div 2 = 9.5\,m^2$

The area is $74.75 + 9.5 = 84.25\ m^2$.

In tolerance notation the area is $(84.25 \pm 9.5)\,m^2$.

b The tub of fertiliser has:

- a minimum possible weight of $(10 - 0.3)\ kg = 9.7\,kg = 9700\,g$

- a maximum possible weight of $(10 + 0.3)\ kg = 10.3\,kg = 10\,300\,g$

  The least possible amount of fertiliser needed is: $74.75 \times 110\,g = 8222.5\,g$

  > Work out the least possible amount of fertiliser needed, i.e. for the smallest possible area of $74.75\,m^2$.

  The greatest possible amount of fertiliser needed is: $93.75 \times 110\,g = 10\,312.5\,g$

  > Work out the greatest possible amount of fertiliser needed, i.e. for the greatest possible area of $93.75\,m^2$.

  This means that if the tub of fertiliser contains:

- the minimum possible weight of $9700\,g$, he will not have enough if his vegetable patch has the maximum possible area

- the nominal weight of $10\,000\,g$, he will not have enough if his vegetable patch has the maximum possible area

- the maximum possible weight of $10\,300\,g$, there will be (almost) enough for the largest possible area.

  One tub will probably be enough as the vegetable patch is probably not the absolute maximum area and the tub will probably not contain the minimum possible weight of fertiliser.

  Although the tolerances tell us that he may not have enough, common sense tells us that it is not necessary to buy more than one tub as the actual measurements will be somewhere between the least and the greatest.

Example 16.7

Andrea is planning to drive from Inverness to Perth.

The distance is (110 ± 5) miles and her satnav tells her to expect to travel at an average speed of (45 ± 0.3) mph.

What are the shortest and the longest journey times that Andrea can expect?

$T = D \div S$

> Write the formula to calculate time.

The shortest possible journey time will be $105.5 \div 45.3 = 2.329...\,h = 2\,h\ 20\,min$.

> To find the longest possible journey time, calculate the longest possible distance divided by the least possible speed.

The longest possible journey time will be $110.5 \div 44.7 = 2.472...\,h = 2\,h\ 28\,min$.

> To find the shortest possible journey time, calculate the shortest possible distance divided by the greatest possible speed.

Exercise 16C

1 Five boys are measured to have a mass of 50 kg, 40 kg, 42 kg, 63 kg and 72 kg with a tolerance of ± 2 kg. Calculate the least and greatest possible total mass of the boys.

2 Sara records her six best times for running 400 m. Each time is given to the nearest 0.1 s. The recorded times are:

56.4, 57.2, 53.9, 57.6, 54.1, 52.9

If her mean time is no more than 55.35 s, she will qualify for a grant to help to pay for her training.

When she applies for the grant, she is turned down. Why does she not qualify for the grant? Use your working to justify your answer.

★ 3 A maths book has a mass of (535 ± 10) g.

   a What is the greatest possible mass of eight of these books?

   b A bookshelf can safely hold up to (8 ± 0.5) kg. How many maths books can safely be put on the shelf?

4 A machine cuts lengths of rope from a 50 m roll. The lengths are $(3.4\text{m} \pm 2\%)$.

What are the minimum and maximum numbers of pieces of rope that can be cut from the 50 m roll?

★ 5 Alistair is training for a 5 km fun run. He runs around the edge of a football pitch with dimensions as shown.

How many times should Alistair run round the pitch so that he runs at least 5 km?

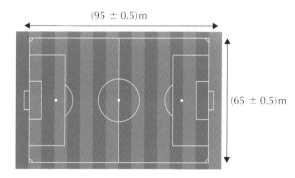

(95 ± 0.5)m

(65 ± 0.5)m

6 A rectangular notice board has length (80 ± 0.5) cm and breadth (40 ± 0.5) cm. Calculate the greatest possible area of the notice board.

7 A shoe box has these dimensions:

   • length (28.6 ± 0.1) cm

   • breadth (19.3 ± 0.1) cm

   • height (9.9 ± 0.1) cm.

What are the minimum and maximum possible volumes of the box? Give your answers in tolerance notation.

8 Harry is going to paint the wall of his bedroom.

He measures the wall to be 4.6 m by 3.3 m to the nearest 0.1 m.

The paint that he wants to use is available in (2.5 ± 0.1) litre tins.

The label on the tin of paint claims that 2.5 litres of paint will cover 30 m².

He is going to put two coats of paint on the wall.

Will one tin of paint definitely be enough for him to finish painting his wall? Use your working to justify your answer.

★ **9** Mrs Shannon's oil tank for central heating measures 180 cm by 130 cm by 120 cm. Each of these measurements is given to the nearest cm.

When the tank is 25% full, Mrs Shannon orders extra oil. What is the maximum number of litres that she should order?

10 Squeezyfruit Drinks produces a carton of juice in the shape of a cuboid. The carton has dimensions (6.9 ± 0.2) cm by (3.9 ± 0.2) cm by (18.6 ± 0.2) cm.

    a The company claims that there is at least 500 m*l* of juice in each carton. Is this claim justified?

    b The company makes a batch of 5000 litres of juice to fill the cartons. Calculate the minimum and maximum possible numbers of cartons that can be filled with this volume of juice.

    c The company makes a profit of £0.36 on each carton of juice that they sell. What are the minimum and maximum profits that can be made from this batch of juice?

✿ ★ **11** Fiona wants to put new tiles on her bathroom floor.

She wants to buy square tiles with a side of length (30 ± 0.1) cm.

    a What are the minimum and maximum possible areas of each tile? Give your answer in tolerance form.

    b Fiona's floor measures exactly 3.2 m by 1.5 m. Calculate how many tiles Fiona will need to cover her floor.

    c Fiona's partner would prefer to use rectangular tiles measuring (20 ± 0.1) cm by (40 ± 0.1) cm. How many of these rectangular tiles will be required?

    d The square tiles are priced at £35 for a pack of 6. The rectangular tiles are priced at £47.50 for a pack of 10. Which shape of tile will be cheaper for Fiona to use?

12 The density of gold is 19.3 g/cm³. A solid gold bar is in the shape of a cuboid with dimensions (4.6 ± 0.05) cm, (6.6 ± 0.05) cm and (2.2 ± 0.05) cm.

    a What are the maximum and minimum possible volumes of this bar of gold? Give your answers to 4 significant figures.

    b What are the maximum and minimum possible values for the mass of this bar of gold?

    c The price of pure gold is £24.09 per gram. What are the minimum and maximum possible values of the gold?

★ **13** A helicopter flying from Aberdeen to two oil rigs before returning home covers a distance of (255 ± 5) km. The helicopter flies at an average speed of $(120 \text{ km/h} \pm 3\%)$. Calculate the minimum and maximum possible time taken to complete the whole journey.

14 A hydraulic jack is used to lift and stabilise a crane platform. The jack stands on a plate measuring 0.3 m by 0.4 m, both ± 0.05 m. During a lift the site foreman calculates the force exerted on the jack as 392 000 N ± 15 000 N.

The pressure exerted by the jack is calculated using this formula:

$$\text{pressure} = \frac{\text{force}}{\text{area}}$$

where pressure is measured in Pascals (Pa), force is measured in newtons (N) and area is in m².

> Look back at Chapter 10 for the relationships between volume, density and mass.

If the pressure exerted by the jack exceeds 4.6 million Pa it will fail. Is the jack safe to complete this lift?

15 A stopwatch records the times for the winner of a 100 metre race as (10.4 ± 0.05) seconds.

   a What are the greatest and least possible times for the winner?

       The length of the track is (100 ± 0.05) metres.

   b Calculate the slowest and fastest possible average speeds of the winner. Give your answers correct to 2 decimal places.

GO! Activity

In supermarkets, pre-packed products are sold in packages by weight or by measure.

To meet Trading Standards legislation, the label on the packaging must state either the minimum quantity or the average quantity that the package contains.

If the minimum quantity is stated, the package must contain at least the advertised amount.

If any package is found to contain less than the advertised minimum, the supermarket could be prosecuted for not complying with the regulations.

For average quantity labelling, the rules are that, for a random sample of the packages:

* the average (mean) contents of the sample must be at least the nominal quantity

* no more than 1 in 40 or 2.5% of the sample may be non-standard, where a non-standard package is one that contains less than the nominal quantity minus the tolerable negative error. The tolerable negative error can be calculated using the table shown below.

* no package in the sample may be inadequate. An inadequate package is one that is less than the nominal quantity minus twice the tolerable negative error.

Nominal quantity (g or m*l*)	Tolerable negative error
5–49	9% of nominal quantity
50–99	4.5 g or m*l*
100–199	4.5% of nominal quantity
200–299	9 g or m*l*
300–499	3% of nominal quantity
500–999	15 g or m*l*
1000–9999	1.5% of nominal quantity
10 000–14 999	150 g or m*l*
15 000 and above	1% of nominal quantity

(continued)

1   Use the table to answer the following questions. In each case the items are labelled as containing average quantity.

    a   Cartons of orange juice are labelled as containing 225 m*l*. The final carton in a sample of 40 is found to contain only 217 m*l*. Is this a non-standard carton?

    b   A bottle of fabric conditioner is labelled as 1.5 litres. When measured it contains 1.48 litres. Is this acceptable?

    c   A tin of tomatoes is labelled as 400 g. When checked, what values would make this:

        i   a non-standard package   **ii**   an inadequate package?

2   At home, take a look at the variety of packages or products that you can find in the cupboards. Each will be labelled with its weight or capacity.

    a   Choose some products from your cupboard and weigh them or count the contents.

    b   Make a table of your results.

    c   Decide if any of the products in your list are non-standard or inadequate.

- I can calculate the limits of measurements given different levels of tolerance. ★ Exercise 16A Q1, Q9

- I can explain the use of appropriate levels of tolerance for fitting components. ★ Exercise 16B Q2, Q4

- I can solve problems involving tolerance calculations. ★ Exercise 16C Q3, Q5, Q11, Q13

For further assessment opportunities, see the Case Studies for Unit 2 on pages 255–257.

17 Investigating a situation involving a gradient

This chapter will show you how to:

- investigate a situation involving a gradient using vertical over horizontal distance
- use coordinates when investigating a situation involving a gradient.

You should already know:

- how to determine the gradient of a slope using vertical height and horizontal distance
- how to use your knowledge of the coordinate system to plot and describe the location of a point on a grid (including a four-quadrant grid)
- how to find the length of a side of a right-angled triangle using Pythagoras' theorem.

Investigate a situation involving a gradient

The **gradient** of a slope is a measure of the steepness of the slope. It is found by comparing vertical height and horizontal distance.

Since height can be thought of as a vertical distance, we can calculate the gradient, m, using the following formula:

$$\text{Gradient} = \frac{\text{vertical distance}}{\text{horizontal distance}}$$

It is sometimes written as:

$$m = \frac{\text{rise}}{\text{run}}$$

where rise is the vertical distance and run is the horizontal distance.

The slanted line is **steeper** in the picture on the right than the one on the left. A steeper line means a bigger gradient, as more height is gained over the same horizontal distance.

STEP BACK IN TIME

The letter m is used to represent the gradient of a line. There are a number of possible reasons for this.

- *It is abbreviated from modulus and comes from the Latin modus which means 'measure'.*
- *It is abbreviated from monter which means 'to climb' in French.*
- *Its earliest known use was in an 1844 British text by Matthew O'Brien entitled 'A Treatise on Plane Coordinate Geometry'. The use of m for gradient may have been chosen at random!*

Gradient can be represented as:

- a fraction (in its simplest form)
- a decimal fraction
- a percentage
- a ratio (vertical : horizontal).

Fractions and decimal fractions are usually used to describe the gradients of lines on a graph. Decimal fractions and percentages are useful when comparing different gradients, or when comparing a gradient to a recommended guideline, but they are not always exact. In real life, gradient is often quoted as a ratio.

For example, a gradient of $\frac{1}{5}$ could also be written as 0.2, 20% or 1 : 5 (sometimes written as 1 in 5).

This chapter uses gradients to compare lines with each other and with a recommended guideline. In any question involving comparison, you will need to give a reason for your decision.

Example 17.1

A community centre needs to build a ramp for wheelchair access into the building.

The local council's requirements are:

- maximum gradient for ramps and paths less than 5 m in horizontal length is 1 : 15
- maximum gradient for ramps and paths more than 5 m in horizontal length is 1 : 20.

The council has two possible designs.

Design 1   **Design 2**

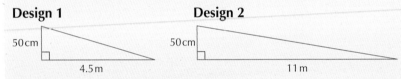

50 cm   4.5 m

50 cm   11 m

Which ramp would you recommend?

Design 1

$$m = \frac{50}{450} = \frac{1}{9}$$

Make sure that the horizontal and vertical distances are in the same units. 4.5 m = 450 cm

$$= 0.111 \text{ (3 d.p.)}$$

Convert to a decimal fraction in order to compare with the standards given.

Maximum gradient allowed for ramp less than 5 m is $\frac{1}{15} = 0.067$ (3 d.p.)

So design 1 does not meet the requirements as 0.111 is greater than 0.067.

In this example you don't need to convert $\frac{1}{9}$ and $\frac{1}{22}$ to decimal fractions to compare them, as it is obvious that $\frac{1}{9}$ is more than $\frac{1}{22}$. However, converting to decimal fractions helps you to compare with the guidelines and solve the problem.

Design 2

$$m = \frac{50}{1100} = \frac{1}{22}$$

$$= 0.045 \text{ (3 d.p.)}$$

Maximum gradient allowed for ramp more than 5 m is $\frac{1}{20} = 0.05$ (2 d.p.)

So design 2 meets the requirements as 0.045 is less than 0.05.

Design 2 is recommended because it meets the council's requirements whereas design 1 does not meet the requirements.

Parallel gradients

Lines which have the same gradient are **parallel**.

For example, compare these two slopes.

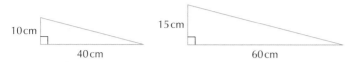

The gradient of the triangle on the left is:

$$m = \frac{10}{40} = \frac{1}{4}$$

The gradient of the triangle on the right is:

$$m = \frac{15}{60} = \frac{1}{4}$$

These two slopes have the same gradient so they are parallel.

Gradient of a horizontal line

A **horizontal** line has a gradient of zero. This is because there is no vertical change.

The calculation of the gradient of a horizontal line will look like this:

$$m = \frac{0}{\textbf{horizontal distance}} = 0$$

Gradient of a vertical line

A vertical line has an **undefined** gradient. This is because there is no horizontal change.

The calculation of the gradient of a vertical line will look like this:

$$m = \frac{\textbf{vertical distance}}{0} = \textbf{undefined}$$

Exercise 17A

1   A cyclist rates a hill (cycling upwards) in terms of the steepness of its gradient like this:

- 0–3%: easy
- 3–6%: manageable
- 6–9%: slightly challenging
- 9–15%: very challenging
- more than 15%: painful.

Using these ratings, describe the following hills, giving a reason for your answer.

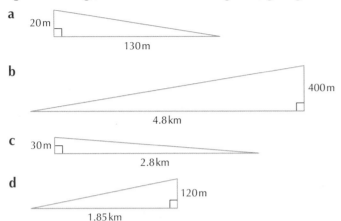

a 20 m — 130 m

b 400 m — 4.8 km

c 30 m — 2.8 km

d 120 m — 1.85 km

2   Building regulations require a roof to have a minimum gradient of 0.27.

The following roofs are isosceles triangles.

Do the following roofs meet the requirement?

Give a reason for each answer.

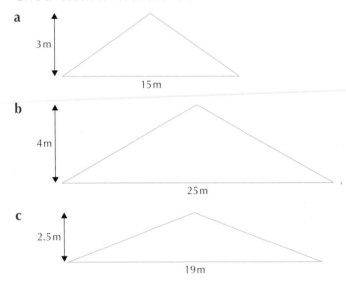

a   3 m — 15 m

b   4 m — 25 m

c   2.5 m — 19 m

3   a   One of the world's most dangerous roads is the Yungas Road in Bolivia, South America, where more than 200 people are killed each year.

A part of the road gains 215 m in height over a horizontal distance of 500 m. Calculate the gradient of this part of the road.

b   Part of the road 'La Route des Crêtes' near Cassis in France has a sign warning that the gradient is 30%. Is this road steeper than Yungas Road? Give a reason for your answer.

4   a   At Ayton Airport the safe gradient for a plane taking off is between 0.05 and 0.1. If a plane leaves Ayton Airport and is recorded at a height of 50 m when the plane is a horizontal distance of 400 m away from the airport, did it take off safely? Give a reason for your answer.

b   The safe gradient for a plane landing is between 2.1% and 2.5%.

A plane lands with a gradient of 2.5%. If it is recorded at a horizontal distance of 500 m from the airport, what is its vertical height at this point?

c Why is the take-off gradient different to the landing gradient?

5 Jonny is a window cleaner. He places his ladder against a wall. The ladder is 9 m long.

a He places the ladder so that the base is 2 m horizontally from the wall.

Use Pythagoras' theorem to calculate the height of the wall reached by the top of the ladder.

b Use your answer to part **a** to find the gradient of the ladder.

c The Health and Safety Executive (HSE) state that the maximum gradient of a ladder placed against a wall is 4 in 1 (that is, 4 units up for every 1 unit on the ground).

Has Jonny put the ladder far enough out from the wall to meet the guidelines? Give a reason for your answer.

d If he is to meet the requirement exactly, how far away from the wall should he place the ladder?

★ 6 The Department of Health's recommendations for stairs, ramps and escalators states the following:

The recommended maximum gradient should be 0.58 with maximum speed 0.75 m/s. A gradient of 0.7 is permitted for vertical rises less than 6 m and where the speed is less than 0.5 m/s.

Do the following meet the recommended guidelines? Give a reason for each answer.

a

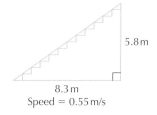

5.8 m
8.3 m
Speed = 0.55 m/s

b

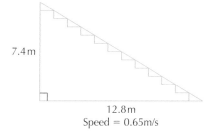

7.4 m
12.8 m
Speed = 0.65 m/s

c

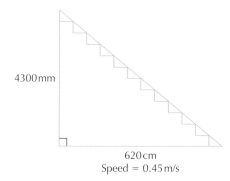

4300 mm
620 cm
Speed = 0.45 m/s

d

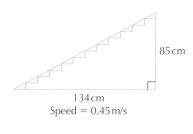

85 cm
134 cm
Speed = 0.45 m/s

7   Tennis courts are built on a slight incline to assist drainage after rainfall.

The recommended gradients for porous courts are between 1 : 200 and 1 : 250.

The recommended gradients for non-porous courts are between 1 : 100 and 1 : 120.

Do these tennis courts meet the recommended guidelines? Give a reason for your answer.

a   Hard court (porous)

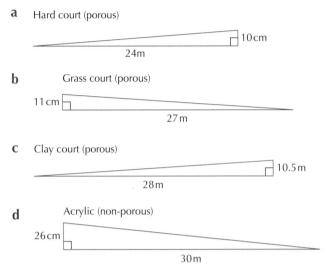

10 cm

24 m

b   Grass court (porous)

11 cm

27 m

c   Clay court (porous)

10.5 m

28 m

d   Acrylic (non-porous)

26 cm

30 m

🔅 Making the Link

A porous tennis court surface uses material made full of tiny holes (called pores) which the water can drain through. Examples of this are hard court, grass, synthetic grass and synthetic clay surface.

A non-porous tennis court surface is an artificial surface and does not use material full of holes. It is usually acrylic and is either used as an inside court or requires regular removal of excess water.

---

8   The Aberystwyth Cliff Railway is a funicular railway with a rise of 130 m and a run of 237 m.

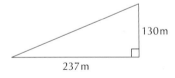

130 m

237 m

a   Calculate the gradient of the funicular.

The Cairngorm Funicular Railway has a rise of 462 m and is 1970 m in length.

1970 m

462 m

b   **i**   Using Pythagoras' theorem, calculate the horizontal distance (run) of the Cairngorm Funicular Railway.

ii   Using your answer to part **i**, calculate the gradient of the Cairngorm Funicular Railway.

c Which is steeper: the Cairngorm Funicular Railway or the Aberystwyth Cliff Railway? Give a reason for your answer.

d Which is longer: the Cairngorm Funicular Railway or the Aberystwyth Cliff Railway? Give a reason for your answer.

9 a A local council's regulations state that a grass area with a gradient of 0.7 or more cannot be cut with a tractor flail mower (a tractor with attachments so that it can cut rough grass). Which of these grass areas can be cut with a tractor flail mower and which would require an alternative arrangement?

i
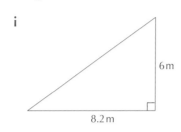

6 m

8.2 m

ii
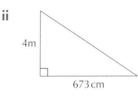

4 m

673 cm

iii
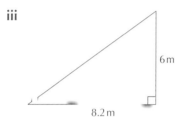

6 m

8.2 m

iv
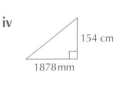

154 cm

1878 mm

b What could the council gardeners do to grass areas which they cannot cut using a tractor flail mower?

10 The Government's document *Protection from Falling, Collision and Impact* specifies that a staircase in a house must follow these regulations:

- the rise of each stair must be between 150 mm and 225 mm

- the run (or tread) of each stair must be between 245 mm and 260 mm

- the gradient must be no more than 0.9.

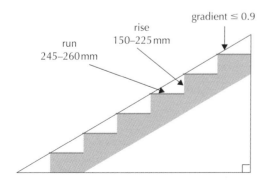

gradient ≤ 0.9

rise
150–225 mm

run
245–260 mm

Does each set of stairs meet the regulations? Give a reason for your answer.

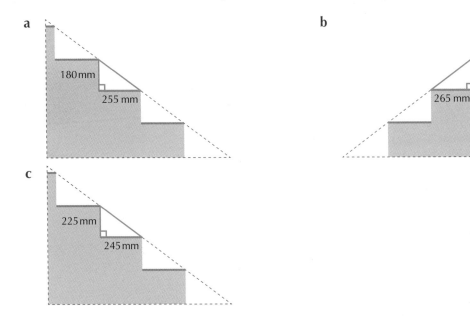

a

180 mm

255 mm

b

190 mm

265 mm

c

225 mm

245 mm

Calculating gradient using a coordinate grid

The gradient of a line on a coordinate grid can be calculated using the formula:

$$\text{Gradient} = \frac{\textbf{vertical distance}}{\textbf{horizontal distance}}$$

This can be worked out by finding the change in vertical position and dividing this by the change in horizontal position. So the formula above can be shortened to:

$$m = \frac{\textbf{vertical}}{\textbf{horizontal}}$$

You can use the squares on a grid to calculate the vertical and horizontal change by creating a right-angled triangle between two points on the line and counting the squares up/down and across.

The gradient of a line on a coordinate grid is usually written as a whole number, a fully simplified fraction or an exact decimal fraction.

- A line sloping upwards from left to right has a **positive** gradient value.

- A line sloping downwards from left to right has a **negative** gradient value (vertical change is downwards so treat it as a negative value).

Example 17.2

Calculate the gradient of each line.

a 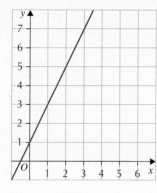 **b**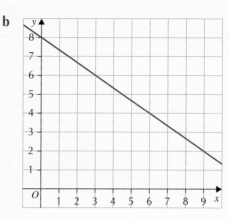

a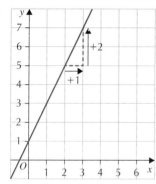

Choose two points on the line and create a right-angled triangle.

$$m = \frac{\text{vertical}}{\text{horizontal}}$$

$$= \frac{2}{1} = 2$$

b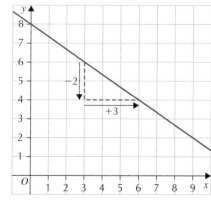

Again, choose two points on the line and create a right-angled triangle. The vertical change is −2 as you are moving **down** the y-axis.

$$m = \frac{\text{vertical}}{\text{horizontal}}$$

$$= \frac{-2}{3} = -\frac{2}{3}$$

Gradient formula

The process of calculating a gradient can be generalised by choosing two points on line AB where A is (x_1, y_1) and B is (x_2, y_2).

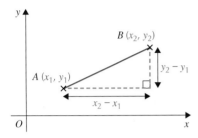

As this diagram shows, if you know two points on a line, then you can calculate the gradient:

$$m_{AB} = \frac{y_2 - y_1}{x_2 - x_1}$$

Example 17.3

Use the formula to calculate the gradients of the lines in Example 17.2.

a $m = \dfrac{y_2 - y_1}{x_2 - x_1} = \dfrac{7 - 5}{3 - 2} = \dfrac{2}{1} = 2$ •———— Use coordinates (2, 5) and (3, 7).

b $m = \dfrac{y_2 - y_1}{x_2 - x_1} = \dfrac{4 - 6}{6 - 3} = \dfrac{-2}{3} = -\dfrac{2}{3}$ •———— Use coordinates (3, 6) and (6, 4).

It is better to use the formula than the diagram of the grid when:

- you are not given a diagram but are asked for the gradient between two coordinates as it is quicker to calculate

- the scales on the horizontal and vertical axes are different.

> ⚠ The calculation of numbers automatically generates a negative value so there is no need to look at the shape of the graph for this purpose.

Exercise 17B

1   A courier delivers packages up to a weight of 11 kg within a city centre.

The courier has three charging bands: packages up to 2.5 kg, packages between 2.5 and 6 kg, and packages over 6 kg. His charge depends on the exact weight of the package.

This graph shows how much he charges.

The courier uses the graph to calculate the charge. For example, a package weighing 3.5 kg would cost £27.

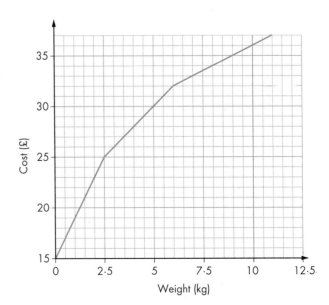

a Use the graph to find the cost of delivering a package weighing:

   i 1.5 kg

   ii 5 kg

   iii 10.5 kg.

b Use the graph to calculate the gradient of each charging band.

c Which size of package gives the largest gradient? Why do you think this is the case?

d What does the gradient indicate in this graph?

e Why do you think that the courier charges different rates according to the weight of the package?

2 Henry was ill in hospital. The graph shows his temperature for the 2 weeks he was in hospital.

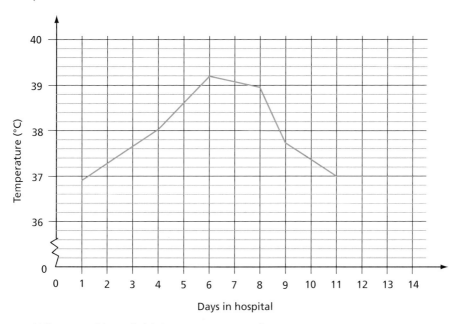

a What was Henry's highest temperature?

b Calculate the gradient of each section of straight line in the graph.

c Between which days did Henry's temperature

   i increase the most

   ii decrease the most?

   Give a reason for each answer.

> Make sure you explain the difference between a positive and negative gradient.

d What does gradient indicate in this graph?

e When Henry's temperature went over 38.5°C, he was put on an antibiotic drip.

   i On which day was he put on an antibiotic drip?

   ii How many days did it take for the antibiotic to work before Henry's temperature started to decrease?

f When Henry's temperature had returned to normal for 4 days he was allowed to go home.

   What is the body's normal temperature? Give a reason for your answer.

g Explain why the gradient of the last section is zero.

★ **3** The graph shows the profile of a mountain running race (sometimes called a fell race).

The horizontal axis shows the distance (in miles) of the race and the vertical axis shows the height (in feet) above sea level.

> ⚠ There are 5280 feet in a mile.

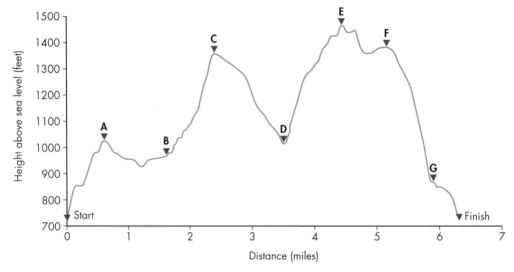

a Calculate the approximate gradient for each section of the graph (between the letters).

b Which section is the steepest hill to run:

  i up

  ii down?

Give a reason for each answer.

c What does gradient indicate on this graph?

d Fell races are categorised by the amount of ascent and distance; in this case feet per mile.

The tables show how races are categorised by ascent and distance.

Ascent	
Category	**Ascent**
A	An average of, at least, 250 feet or more per mile
B	An average of 125–250 feet or more per mile
C	An average of 100–125 feet or more per mile

Distance	
Category	**Distance**
L	More than 12 miles
M	6–12 miles
S	Less than 6 miles

So, for example, an AL race would be over 12 miles long and have (on average) at least 250 feet of ascent per mile

What category is the race above? Give a reason for your answer.

4 The graph shows a short passenger journey made in a London black cab on a Tuesday at 7.20 am.

a Calculate the gradient of the graph for each stage of the journey.

b Between which times of the journey (in minutes) was the fastest part of the journey? Give a reason for your answer.

c What is happening between 1 and 2 minutes into the journey? What do you think could be a reason for this?

d The tariffs for hiring the cab between 6.00 am and 10.00 pm are:

- for the first 300 metres or 60 seconds (whichever is reached first) there is a minimum charge of £2.20

- for each additional 200 metres or 30 seconds (whichever is reached first) or part thereof there is a charge of 20p (provided the fare is less than £15.80)

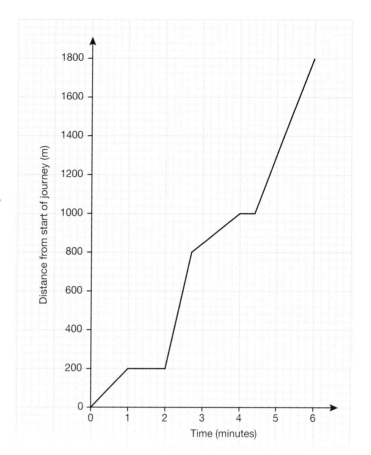

- once the fare is £15.80 or greater, there is a charge of 20p for each additional 100 metres or 21 seconds (whichever is reached first) or part thereof.

  Estimate the fare using this tariff.

e If the taxi is hired between 10.00 pm and 6.00 am the tariffs are different:

- for the first 200 metres or 36 seconds (whichever is reached first) there is a minimum charge of £2.20

- for each additional 100 metres or 12 seconds (whichever is reached first) or part thereof there is a charge of 20p (provided the fare is less than £23.00)

- once the fare is £23.00 or greater, there is a charge of 20p for each additional 98 metres or 21 seconds (whichever is reached first) or part thereof.

  i How much would the journey have cost if the passenger travelled at 10.30 pm?

  ii Calculate the difference in cost between the two tariffs.

  iii Why is there a different hire tariff for different times of the day?

5 The temperature was recorded every hour over a 24-hour period in Perth on a winter's day. The graph shows the results.

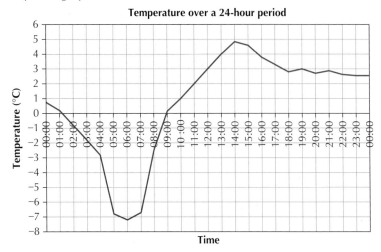

a Calculate the gradient (also known as the rate of change) between:

  i 09:00 and 13:00

  ii 01:00 and 04:00

  iii 16:00 and 18:00.

b What is the maximum rate of change and between which times does it take place? Give a reason for your answer.

c When does the rate of change show little variation? Give a reason for your answer.

> ⚠ The rate of change is the same as the gradient of a graph where time is on the horizontal axis.

6 A small business which sells ice cream records its profit and loss over their first 7 years of business. The graph shows the results.

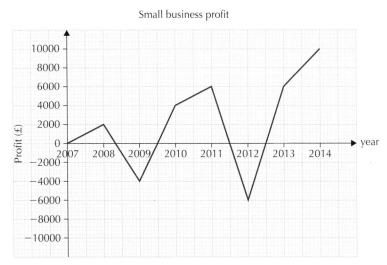

a Calculate the rate of change between:

  i 2008 and 2009

  ii 2011 and 2012

  iii 2013 and 2014.

b Between which years did profit increase the most? Give a reason for your answer.

c Between which years did the greatest loss occur? Give a reason for your answer.

7 A scuba diver's descent is recorded and shown in the graph.

a Calculate the rate of change of the diver's depth between:

   i 0 and 7 minutes

   ii 20 and 30 minutes

   iii 40 and 50 minutes.

b Between which times was she travelling fastest?

c Was she faster on the way down or the way up? Give a reason for your answer.

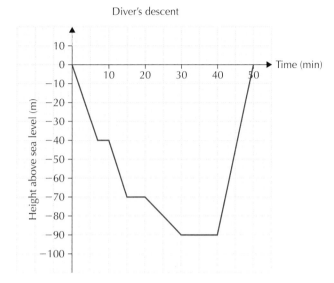

Diver's descent

Activity

1 a Exercise 17A Question 10 refers to the regulations for stairs in a house specified by the Government document *Protection from Falling, Collision and Impact*:

- the rise of each stair must be between 150 mm and 225 mm

- the run (or tread) of each stair must be between 245 mm and 260 mm

- the gradient must be no more than 0.9.

If you have stairs in your home, measure the stairs. Do they comply with the current regulations? Give a reason for your answer.

b Now research the regulations for other staircases (for example, school, stairs outside buildings, shared stairwells/closes), then measure and check whether or not they comply with the current regulations.

- I can investigate a situation involving a gradient using vertical over horizontal distance. ★ Exercise 17A Q6

- I can use coordinates when investigating a situation involving a gradient. ★ Exercise 17B Q3

For further assessment opportunities, see the Case Studies for Unit 2 on pages 255–257.

18 Solving a problem involving a composite shape which includes part of a circle

This chapter will show you how to:

- solve a problem involving a composite shape which includes part of a circle.

You should already know:

- how to find the perimeter and area of simple 2D shapes (e.g. triangle, kite, rhombus, parallelogram, circle)
- how to find the area of a composite shape.

Solving a problem involving a composite shape

A **composite shape** is a 2D shape which can be broken into two or more familiar shapes.

To calculate the **area** of a composite shape, break this shape into smaller parts, calculate the area of each part individually, then add them together.

The table shows how to calculate the area of some familiar shapes.

Name	Shape	Area
Triangle	height, base	**Area = $\frac{1}{2}$ × base × height** $A = \frac{1}{2}bh$
Kite	diagonal 1	**Area = diagonal 1 × diagonal 2** $A = \frac{1}{2}d_1d_2$
Rhombus	diagonal 2	**Area = diagonal 1 × diagonal 2** $A = \frac{1}{2}d_1d_2$
Parallelogram	height, base	**Area = base × height** $A = bh$
Circle	diameter, radius	**Area = π × radius × radius** $A = \pi r^2$

The **perimeter** of a composite shape is the total length around the edge of the shape.

The perimeter of a circle is called the **circumference** and can be calculated using one of these formulae:

$$C = \pi d = 2\pi r$$

where d is the diameter of the circle and r is the radius.

In this chapter you will solve problems about the area and perimeter of composite shapes. All questions will involve a circle or part of a circle.

> The value of π that you use will alter the accuracy of your calculation.

Example 18.1

A window needs to be replaced. It is a composite shape made from a rectangle and a semi-circle.

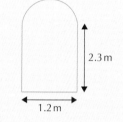

2.3 m

1.2 m

a The window is to be replaced with clear double-glazing which costs £63.50 per square metre. What is the cost of replacing the window?

b Patterned double-glazing is charged at £73.50 per square metre. How much more expensive is this option than clear double-glazing?

c The window currently has an aluminium frame. To upgrade to a steel window frame will cost £22.50 per metre. What is the cost of replacing the window frame?

a Area of rectangle is $1.2 \times 2.3 = 2.76\,\text{m}^2$

Area of circle is $\pi \times 0.6^2 = 1.13097\,\text{m}^2$ ← The radius is half of 1.2 m which forms the diameter of the circle.

Area of semi-circle is $1.13 \div 2 = 0.56548\,\text{m}^2$ ← The area of the semi-circle is half the area of the circle.

Total area is $2.76 + 0.565 = 3.32548\,\text{m}^2$

Cost is $3.32548 \times £63.50 = £211.17$ (2 d.p.)

b Cost of patterned double-glazing is $3.325 \times £73.50 = £244.42$

Difference in cost is $£244.42 - £211.17 = £33.25$

c Circumference of circle is $\pi \times 1.2 = 3.77\,\text{m}$

Perimeter of shape is $\dfrac{3.77}{2} + 2.3 \times 2 + 1.2 = 7.68\,\text{m}$ ← The perimeter of the semi-circle is half the circumference of the circle.

Cost of steel frame is $7.68 \times £22.50 = £172.91$ (2 d.p.)

Area of a sector of a circle

The **sector** of a circle is the area between two radii. It is bounded by the two radii and the **arc** (the part of the circumference between the two radii).

To find the area of a sector, use this formula:

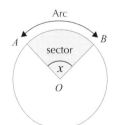

$$\textbf{Area of a sector} = \frac{x}{360} \times \pi r^2$$

To find the length of an arc of a circle, use this formula:

$$\textbf{Arc length} = \frac{x}{360} \times \pi d$$

where x is the angle at the centre of the sector (measured in degrees), r is the radius of the circle and d is the diameter of the circle.

Example 18.2

A lawn consists of two rectangles and a quarter circle.

a Shauna wants to re-turf her lawn. Turf is bought at £2.05 per square metre. How much will this cost her?

b i She has been advised to add lawn soil to the area due to be re-turfed. It is sold in bags of 25 litres and she has been advised to use 1 bag per 2.5 square metres. Only whole bags can be purchased. Bags cost £4.60. How much will this cost her?

ii She has been told that, in autumn, she should add a top dressing of soil using the same lawn soil, but this time at the rate of one 25 litre bag for every 12 square metres of lawn area. How much will this cost her?

c What is the total cost of re-turfing her lawn?

d Shauna wants to put a small fence around the edge of the garden. Fencing costs £2.99 per metre but she can only buy it in complete metres. How much will the fence cost her?

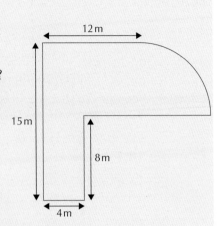

a Area of part 1 is $12 \times (15 - 8) = 12 \times 7 = 84\,\text{m}^2$

Area of part 2 is $\dfrac{90}{360} \times \pi \times 7^2 = 38.48451\,\text{m}^2$

> The breadth of rectangle 1 is 7, so it is also the radius of the quarter-circle.

Area of part 3 is $8 \times 4 = 32\,\text{m}^2$

Total area is $84 + 38.48 + 32 = 154.48451\,\text{m}^2$

Cost of turf is $2.05 \times 154.48 = £316.69$ (2 d.p.)

b i Bags needed is $\dfrac{154.48}{2.5} = 61.79 \approx$

62 bags (rounded up)

Cost of lawn soil is $62 \times £4.60 = £285.20$

ii Bags needed $= \dfrac{154.48}{12} = 12.87 \approx 13$ bags

Cost of lawn soil is $13 \times £4.60 = £59.80$

c Total cost of re-turfing lawn is £316.69 + £285.200 + £59.80 = £661.69

d Circumference of circle $= \pi d = p \times 14 = 43.98\,\text{m}$

Perimeter of shape $= \dfrac{43.98}{4} + 12 + 15 + 4 + 8 + 8 + 7 = 64.995\,\text{m} \approx 65\,\text{m}$ fencing (rounded up to next whole metre)

Cost of fencing is $65 \times £2.99 = £194.35$

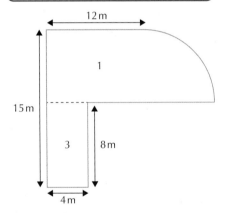

For obvious sector fractions, a simpler fraction can be used instead of $\dfrac{angle}{360}$.

In this example, $\dfrac{90}{360}$ could be replaced with $\dfrac{1}{4}$.

Exercise 18A

1 Dave is painting a wall.

He buys a tin of paint that will cover 10 m².

Does he have enough paint?

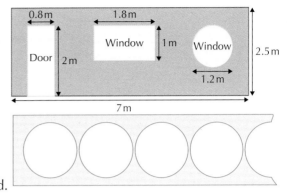

2 Bottle tops are stamped from rectangular metal strips as shown below.

Each bottle top is made from a circle of radius 1.7 cm. Each rectangular strip measures 4 cm by 500 cm. There is a gap of 0.2 cm between each circle and a 0.2 cm of metal needs to be left at the end.

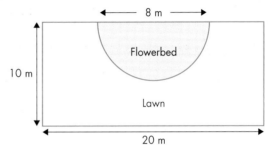

a How many bottle tops can be stamped out of one strip?

b How much metal is wasted when one strip is stamped?

c A new machine has been developed which can stamp out bottle tops without needing to leave a 0.2 cm gap at the end and between each circle. How much less waste is generated using this machine?

d Give two reasons why the new machine should be used.

3 The diagram shows a plan of Mrs Allison's new back garden.

She decides to plant grass seeds to grow a new lawn.

a Her local garden centre sell 500 g packets of grass seed for £8.99.

She is advised that 1 kg of grass seed covers 20 m².

i How many packets does she need to buy?

ii How much will this cost her?

b She decides to check the cost online before purchasing the grass seed, and finds the same brand of grass seed on an internet site for £5.99 per packet. How much money will she save if she buys the packets online?

c She decides to put wire mesh around the flowerbed to protect the flowers. Wire mesh is sold for £1.50 per metre at her local garden centre. How much will the mesh cost her?

d She also checks the cost online before purchasing the wire mesh, and finds the same type of mesh for sale in 5 metre rolls at £6 per roll (only whole rolls can be purchased). Is she better to buy this online or at her local garden centre? Give a reason for your answer.

4 A company has designed a new logo for the outside of its office, showing a circle fitting exactly inside a semi-circle of 1.2 metres.

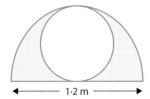

The logo is to be made from plastic. Plastic is sold at 5p per 10 cm².

Calculate the cost of the logo.

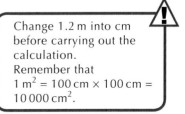

Change 1.2 m into cm before carrying out the calculation.
Remember that
1 m² = 100 cm × 100 cm = 10 000 cm².

5 A company makes tablecloths with lace round the outside. It is a composite shape made from a rectangle and two semi-circles.

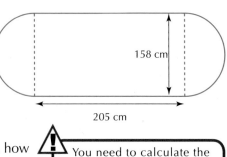

158 cm

205 cm

a The material costs £1.39 per square metre. What is the cost of the material for 1 tablecloth?

b The lace round the edge costs £0.48 per metre. How much does this cost?

c If they sell tablecloths for £12.50 per tablecloth, how much profit do they make on 20 tablecloths?

> ⚠ You need to calculate the perimeter of the shape. To do this you will need to find the circumference of the circle using $C = \pi d$ and add 205 cm × 2.

6 A small bridge is built as a garden feature.

a Calculate the surface area of the front of the bridge.

b The front and back of the bridge are the same. Both surfaces need to be varnished. One 5 litre tin of varnish covers $24\,\text{m}^2$.
Will one tin of varnish be enough? Give a reason for your answer.

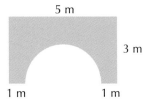

5 m

3 m

1 m   1 m

7 A restaurant serves pizza which is a composite shape made up of two semi-circles and a rectangle with a circle removed from the centre (for salad).

It sells two sizes of pizza.

The smaller pizza has the dimensions shown above and the circle removed has radius 4 cm.

The larger pizza has the dimensions shown on the right and the circle removed has radius 6 cm.

The restaurant claims that the larger size of pizza is exactly 50% bigger than the smaller.

Is this claim correct? Give a reason for your answer.

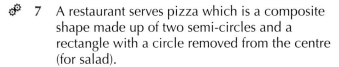

15 cm

4 cm

20 cm

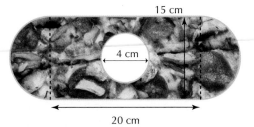

20 cm

6 cm

30 cm

8 A pond is a composite shape made from a rectangle and a semi-circle.

There is a gravel path surrounding the pond which is 1 metre wide.

a What is the area of the gravel path?

b Gravel is bought in packets costing £2.49 each. One packet covers $0.4\,\text{m}^2$. How much does it cost to cover the path in gravel?

c A small fence is to be placed surrounding the pond to prevent gravel going into the pond. The fence costs £1.85 per metre and can only be sold in whole metre lengths. How much will the fence cost?

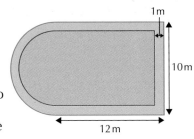

1m

10m

12m

★ **9** Scott's living room is a composite shape consisting of two rectangles with an isosceles triangle removed from one corner. It also has a circle removed from one corner as it has a spiral staircase from the living room to the upstairs of the house. The floor plan is shown below.

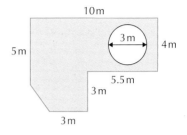

a Scott wants new flooring for his living room. The carpet he likes is sold for £15.99 per square metre, and the shop charges £3.00 per square metre for fitting the carpet. How much will it cost him for a new carpet?

b Laminate flooring costs £22.99 per square metre and the shop charges £10 per square metre for fitting the flooring. How much more will it cost to laminate the floor than to pay for a new carpet?

10 Fred is a joiner and decides to have a new logo painted on his fleet of vans. Smaller logos are cheaper than bigger ones because they use less paint. Which of these logos should he use?

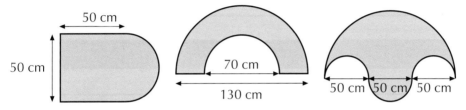

11 'The General Sherman' is a tree in Sequoia National Park in USA and is considered to be the largest in the world. Its base has a diameter of 11.1 metres.

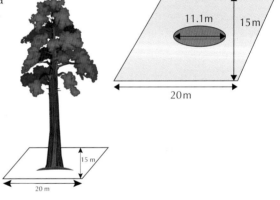

a If the park authority decides to put a fence around the tree in the shape of a parallelogram with the measurements shown, what area of ground (excluding the area taken up by the tree) will the fence enclose?

b The area enclosed by the fence surrounding the tree is to be fertilised with a special fertiliser costing £8.30 for 8 litres. 1 litre of fertiliser should be used for every $20\,m^2$ of ground. How much will fertiliser cost each year if the tree is treated every 4 weeks?

12 The diagram shows a quarter of a circle. The shaded area is to be used to make a woollen rug for placing in front of a fire. The material is sold as 8p per square centimetre. Calculate the cost of the rug.

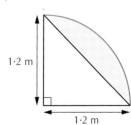

13 ABCD is a square of side length 18 cm. It is part of a design for a home-made birthday card.

   a Calculate the area of the shaded part.

   b The unshaded part is to be covered in silver glitter. 10 g of glitter covers 3 cm² of card. Glitter is sold in 50 g packets and each packet costs 79p. Calculate the cost of covering the unshaded part in silver glitter.

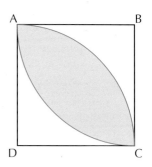

14 A shelf is to be cut in the shape of a quarter circle so it fits in a corner of a room. It is to be cut from a piece of wood measuring 30 cm by 40 cm.

   a What is the maximum possible area of the shelf?

   b What is the area of wood left over?

15 The diagram opposite shows a keyhole shape.

   a Calculate the area of this shape.

   b The keyhole is a hole in a rectangular door measuring 1.2 m by 2.1 m.

   i What is the area of the door?

   ii The door is to be painted red. Red paint is sold in 375 ml pots. One pot of paint covers 5 m². The owner decides to paint the door with two coats of paint. Is one tin of paint enough to do this? Give a reason for your answer.

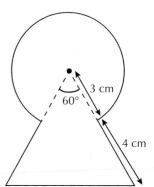

16 The diagram shows the inside of a running track.

   The groundsperson is going to cover this area with grass seed. One 20 kg sack of grass seed covers 325 m² and costs £51.99.

   a How many sacks are needed?

   b How much will this cost?

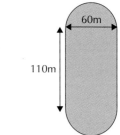

17 A Ferris wheel has a design on both sides of the wheel as shown.

   The diameter of the wheel is 25 m and the length of the base of each isosceles triangle is 9 m.

   a Using Pythagoras' theorem, calculate the altitude of each triangle.

   b Calculate the area coloured:

   i blue

   ii green

   iii red.

   Tins of paint are designed to cover an area of 12 m².

   Red paint costs £12.99, green paint costs £14.99 and blue paints costs £15.99 per tin.

   How much does it cost to paint the Ferris wheel?

18 A running track has the dimensions shown.

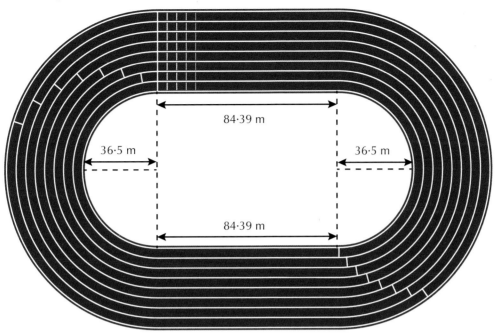

Between each lane there is exactly 1.22 m. There are 8 lanes. The track is to be outlined in white paint.

What is the length of track that is to be painted white?

⊙ GO! Activity

The diagram shows a square within a circle and a square outside the same circle.

There are three coloured regions in this shape: the blue region, the green region and the white region.

Which region has the largest area? Which region has the smallest area?

Justify your answer.

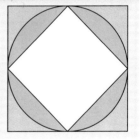

* I can solve a problem involving a composite shape which includes part of a circle. ★ Exercise 18A Q9

For further assessment opportunities, see the Case Studies for Unit 2 on pages 255–257.

a

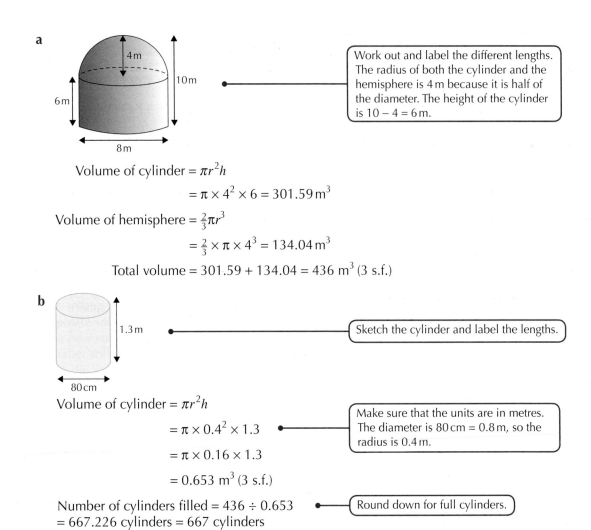

Work out and label the different lengths. The radius of both the cylinder and the hemisphere is 4 m because it is half of the diameter. The height of the cylinder is 10 − 4 = 6 m.

Volume of cylinder = $\pi r^2 h$

$$= \pi \times 4^2 \times 6 = 301.59\,\text{m}^3$$

Volume of hemisphere = $\frac{2}{3}\pi r^3$

$$= \frac{2}{3} \times \pi \times 4^3 = 134.04\,\text{m}^3$$

Total volume = $301.59 + 134.04 = 436\,\text{m}^3$ (3 s.f.)

b

Sketch the cylinder and label the lengths.

Volume of cylinder = $\pi r^2 h$

$$= \pi \times 0.4^2 \times 1.3$$

Make sure that the units are in metres. The diameter is 80 cm = 0.8 m, so the radius is 0.4 m.

$$= \pi \times 0.16 \times 1.3$$

$$= 0.653\,\text{m}^3 \text{ (3 s.f.)}$$

Number of cylinders filled = $436 \div 0.653$

Round down for full cylinders.

$= 667.226$ cylinders = 667 cylinders

Exercise 19A

1 a These three solid objects have been made of different metals. Calculate the mass of each solid object correct to the nearest gram.

i

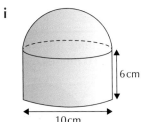

Silver solid object made from a cylinder and hemisphere.

1 cm^3 of silver weighs 10.5 g.

ii

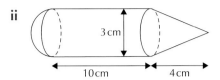

Bronze solid object made from a cylinder, hemisphere and cone.

$1\,cm^3$ of bronze weighs $8.5\,g$.

iii

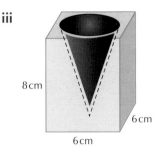

Gold solid object made from a cuboid with a conical section (cone) removed from the cuboid.

The cone removed has diameter $5\,cm$ and height $7\,cm$.

$1\,cm^3$ of gold weighs $19.3\,g$.

  b Which is the heaviest solid object? Give a reason for your answer.

2 A spinning top consists of a cone whose base has a radius $6\,cm$ and has a slant height of $10\,cm$.

  a Calculate the height of the cone.

  b Calculate the volume of the spinning top. Give your answer correct to 3 significant figures.

  c A cylinder has the same volume as the spinning top. Its height is $12\,cm$. What is its radius? Give your answer to the nearest millimetre.

> Draw a sketch of a right-angled triangle to represent the slant height, radius and vertical height.

3 An artist is planning to build a solid obelisk composed of a cuboid and rectangular based pyramid, as shown in the diagram.

  a Calculate the volume of the obelisk

  b The artist is planning on using concrete for the obelisk. The concrete is delivered by trucks in batches of $6m^3$, How many deliveries are required to complete the project?

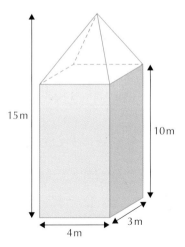

★ **4 a** The diagram shows a laboratory flask made from a sphere with a cylindrical neck.

The flask is filled so that it leaves exactly 1 cm at the top. Calculate the volume of liquid that can be poured into the flask. Express your answer correct to the nearest millilitre.

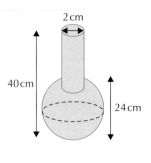

b A laboratory test tube is a composite object made of a cylinder and hemisphere.

The cylinder has a diameter of 3 cm and the total height of the test tube is 15 cm.

Calculate the volume of liquid that can be poured into the test tube if a gap of 1 cm is left at the top of the test tube. Express your answer correct to the nearest millilitre.

c i How many test tubes can be filled from the flask?

ii How much liquid is left over?

5 A swimming pool is a composite object which consists of a cuboid and half a cylinder.

How many litres are needed to fill the swimming pool to a depth of 120 cm?

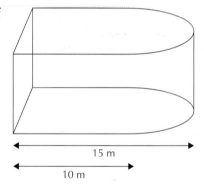

6 A frustum is the remainder of a regular solid object whose upper part has been cut off by a plane parallel to the base.

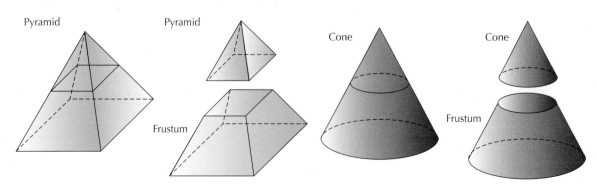

Pyramid   Pyramid   Cone   Cone

Frustum   Frustum

A frustum is made from pyramid A by removing pyramid B.

Pyramid *A*

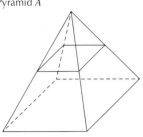

Pyramid *B*

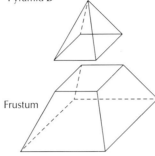

Frustum

The base of pyramid A is a square of side 8 cm and its height is 12 cm.

The base of pyramid B is a square of side 3 cm and its height is 4.5 cm

a Calculate the volume of the frustum.

b Express the volume of pyramid B as a percentage of the frustum.

c 1 cm^3 of mahogany wood weighs 0.67 g. What is the mass of the frustum if it is made of mahogany wood?

7 The base of a new trophy is to be a frustum made from part of a cone.

a Calculate the volume of the trophy base.

b What fraction of the volume of the whole cone was removed?

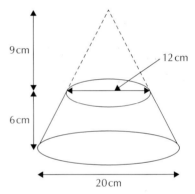

9 cm

12 cm

6 cm

20 cm

8 Helena is playing with modelling clay and makes this solid object which consists of a cone and hemisphere.

a What is the volume of the solid object?

b Helena takes the modelling clay and re-shapes it into a sphere. Calculate the radius of the sphere, correct to 2 significant figures.

10 cm

14 cm

9 A tennis ball of radius 3.5 cm is packaged into a cylindrical box. The ball touches the sides, top and base of the box.

What fraction of the volume of the box is empty space?

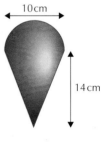

3.5 cm

10 The diagram shows an oil container which consists of a cylinder and a hemisphere.

The oil takes up half of the volume of the container.

Find the depth of the oil which is marked *x*.

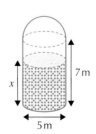

7 m

x

5 m

11 An ice cream carton is a composite object consisting of a cuboid and two half-cylinders.

Ice cream is scooped out of the carton using an ice cream scoop which produces spheres of diameter 4 cm.

How many scoops can you get from 5 tubs of ice cream?

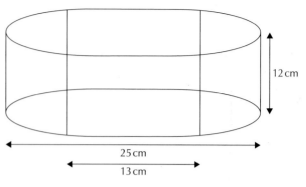

12 cm

25 cm

13 cm

12 Three golf balls of diameter 43 mm just fit inside a cylindrical container.

  a What is the volume of empty space in the container?

  b What percentage of the container is empty?

13 The diagram shows a solid object which consists of a cuboid and pyramid.

The cuboid and the pyramid have the same height.

The total volume of the solid object is 160 cm^3.

Calculate the height of the solid object.

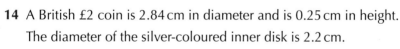

3 cm

8 cm

14 A British £2 coin is 2.84 cm in diameter and is 0.25 cm in height.

The diameter of the silver-coloured inner disk is 2.2 cm.

  a Calculate the volume of the gold-coloured outer ring.

  The silver-coloured inner disk is made of cupronickel. 1 cm^3 of cupronickel weighs 8.95 g.

  b The gold-coloured outer ring is made of nickel brass. 1 cm^3 of nickel brass weighs 8.4 g.

  Calculate the weight of a £2 coin.

15 Spheres of radius 3 cm are packed into a cylinder.

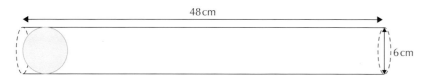
48 cm

6 cm

    a   How many spheres will fit into the cylinder?

    b   When the cylinder is filled with the maximum number of spheres it can hold, the cylinder is filled with water. Calculate the volume of water in the cylinder.

       The spheres are made of aluminium. 1 cm^3 of aluminium weighs 2.7 g. 1 ml of water weighs 1 g.

    c   Calculate the weight of the contents of the cylinder.

    d   The aluminium spheres are replaced with zinc spheres. 1 cm^3 of zinc weighs 7.14 g.

       How much heavier is the cylinder now?

16 The cylinder and sphere have the same volume and radius.

The diameter of the sphere is 12 cm. What is the height of the cylinder? Give your answer correct to 1 decimal place.

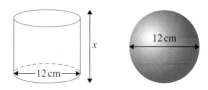
x

12 cm

12 cm

17 A candle consists of a cuboid and pyramid.

    a   Calculate the volume of the candle.

    b   The candle is melted down and made into a cylinder with radius 6 cm. What is the height of the new candle?

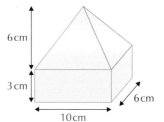
6 cm

3 cm

6 cm

10 cm

18 A large triangular prism is made of chocolate.

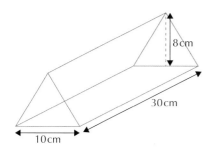
8 cm

30 cm

10 cm

It is melted down and made into small chocolate coins.

Each coin has a diameter of 2.4 cm and a height of 0.3 cm.

How many coins can be made from the triangular prism?

19 A water container consists of a cylinder and hemisphere.

The container is full of water.

Next to the container there are paper cones with diameter 7 cm and height 12 cm.

How many cones of water can be filled from the water container?

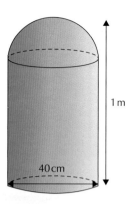

1 m

40 cm

🟢 Activity

a A cone of height 10 cm is cut in two. The small cone and the frustum both have a height of 5 cm.

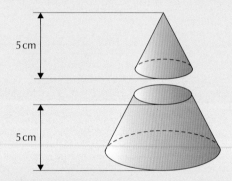

5 cm

5 cm

Explain why the ratio of the volume of the small cone to the frustum is 1 : 7.

⚠️ Consider scale factor.

b Using the above information, calculate the volume of the frustum shown below.

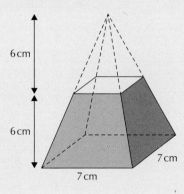

6 cm

6 cm

7 cm

7 cm

• I can solve a problem involving the volume of a composite solid object. ★ Exercise 19A Q4

○ ○ ⬭

For further assessment opportunities, see the Case Studies for Unit 2 on pages 255–257.

20 Using Pythagoras' theorem within a two-stage calculation

Applying Pythagoras' theorem within a two-stage calculation

Pythagoras' theorem is used to find the length of an unknown side in a simple right-angled triangle. Pythagoras' theorem states:

$$a^2 + b^2 = c^2$$

where c represents the hypotenuse (longer side) and a and b represent the shorter sides.

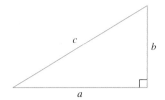

 The formula for Pythagoras' theorem is given at the front of the Lifeskills exam paper.

You can use Pythagoras' theorem in more complicated problems in which right-angled triangles can be constructed. In this chapter you will need to use Pythagoras' theorem more than once in a problem or you may need to use Pythagoras' theorem as well as a different type of calculation in order to solve a problem.

Example 20.1

A child has a set of shapes he can fit together. He found that the red and blue right-angled triangles fitted together exactly along line *BD*.

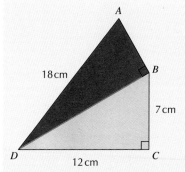

a Calculate the length of *AB*.

b Calculate the perimeter of the whole shape.

(*continued*)

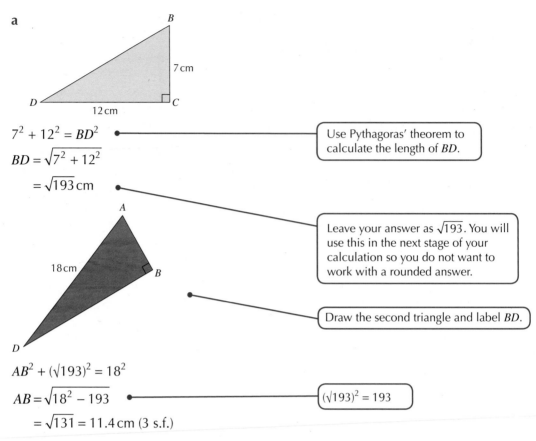

a

$7^2 + 12^2 = BD^2$ •————————— Use Pythagoras' theorem to calculate the length of BD.

$BD = \sqrt{7^2 + 12^2}$

$\quad = \sqrt{193}$ cm   •————— Leave your answer as $\sqrt{193}$. You will use this in the next stage of your calculation so you do not want to work with a rounded answer.

Draw the second triangle and label BD.

$AB^2 + (\sqrt{193})^2 = 18^2$

$AB = \sqrt{18^2 - 193}$ •————————— $(\sqrt{193})^2 = 193$

$\quad = \sqrt{131} = 11.4$ cm (3 s.f.)

b   Perimeter of whole shape is $18 + 7 + 12 + 11.4 = 48.4$ cm

Pythagoras' theorem in three dimensions

You will be asked to find the length of face diagonals and space diagonals of 3D solid objects.

A **face diagonal** is a diagonal (vertex to vertex) which lies on the face of a 3D object.

A **space diagonal** is a diagonal (vertex to vertex) joining opposite vertices and is inside the shape. In a cube or cuboid, the space diagonal goes through the centre of the shape.

For example in the cuboid shown:

- EG is a face diagonal (blue line)
- AG is a space diagonal (red line).

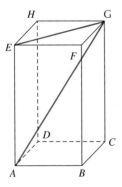

Example 20.2

A set of ornamental chopsticks are 14 cm long and need to be packed in a box.

Will the chopsticks fit into this box?

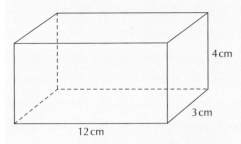

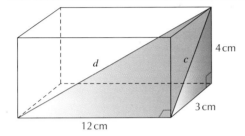

Identify a face diagonal and a space diagonal to help you to solve the question.

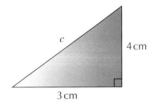

$c^2 = 3^2 + 4^2$

Use Pythagoras' theorem.

$c = \sqrt{3^2 + 4^2}$

$\quad = \sqrt{25} = 5 \text{ cm}$

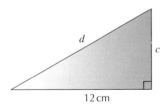

$12^2 + 5^2 = d^2$

$d = \sqrt{12^2 + 5^2} = 13 \text{ cm}$

No, the 14 cm chopsticks will not fit into the box as the space diagonal is only 13 cm which is 1 cm too short.

Exercise 20A

Round your answers to questions in this exercise to 3 significant figures unless stated otherwise.

1   This is the mast of a sailboat.

    a   What is the length of *x* in metres?

    b   What is the length of the perimeter of this mast?

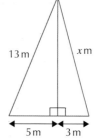

2   A garden consists of two right-angled triangles.

    a   Calculate the length of *x* in metres.

    b   Calculate the area of the garden.

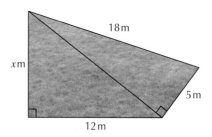

3   A flagpole of length 9 metres is held upright by two guy ropes: one is 15 metres long and the other is 11 metres long. How far apart are the ropes on the ground?

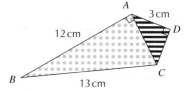

4   Jane uses some material to make a design for her pencil case. The design is made up of two right-angled triangles as shown in the diagram.

    a   Calculate the length of *CD*.

    b   What is the area of Jane's design?

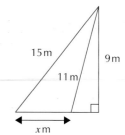

5   A company has designed a new logo for their headed notepaper.

    a   Calculate the lengths of *x*, *y* and *z*.

    b   The company puts a large version of the logo on the outside of their office, using aluminium for each line on the logo. How much aluminium is used?

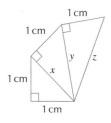

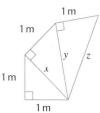

6 Two right-angled triangle-shaped slabs are placed next to each other in a garden in such a way that that they share a common length (*QS*).

Calculate the length of *PQ*, the hypotenuse of the smaller slab.

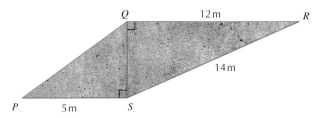

7 A kite has the dimensions shown.

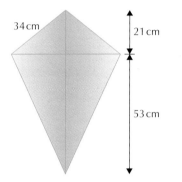

Calculate the total length of wood needed to make the kite frame.

8 Calculate the length of the playground slide shown.

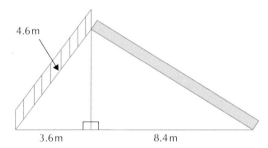

$\oplus^{o}$ **9** The diagram shows a ramp.

The local council specifies that the maximum gradient for ramps and paths under 5 m in horizontal length is 1 : 15.

Does this ramp meet the requirements? Give a reason for your answer.

10 The diagram shows a stained-glass window.

Lead is placed around the outside of the window and between each piece to hold the pieces together. What length of lead is needed for this stained-glassed window?

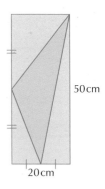

50 cm

20 cm

11 A 7 m ladder is placed against a wall. The bottom of the ladder is 1.5 m from the base of the wall.

a Calculate the height of the wall reached by the top of the ladder.

b Use your answer to part **a** to find the gradient of the ladder.

c The Health and Safety Executive (HSE) states that a ladder must be placed against a wall with a maximum gradient of 4 in 1 (1 unit along the ground for every 4 units up).

Has the ladder been placed far enough out from the wall to meet the HSE guidelines? Give a reason for your answer.

12 A picture is hanging on a string secured to two points at the side of the frame.

Initially the string is 45 cm long. When the picture is hung it stretches as shown in the diagram. By how much does the string stretch?

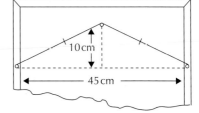

10 cm

45 cm

13 For each of these boxes, calculate the maximum length of wooden rod that can be placed into the box.

⚠ Calculate the length of the space diagonal.

a

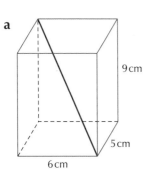

9 cm

5 cm

6 cm

b

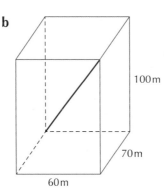

100 m

70 m

60 m

c

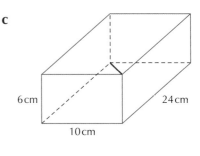

6 cm

10 cm

24 cm

14 Two wooden ornaments are in the shape of pyramids.

   i Calculate the length of the sloping edge on each pyramid.

   ii Calculate the total length of wood required to construct each pyramid.

The dotted lines through each base split the square and rectangle into isosceles triangles. Split one of these isosceles triangles into two equal parts by drawing a line from where the perpendicular line meets the base through to the midpoint of one of the sides of the base.

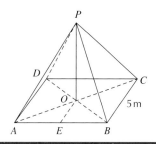

a

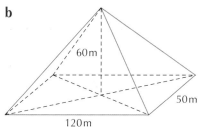

b

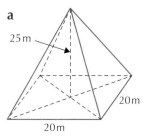

15 A bread tin is made in the shape of a cube with length of side 17 cm. Can a breadstick of length 30 cm fit into the tin with the lid shut?

16 A copper sculpture is made from four equilateral triangles stuck together to make a pyramid.

An art gallery wants to put the sculpture into its main hall which has a vertical height of 7 m. Will the sculpture fit into the space? Give a reason for your answer.

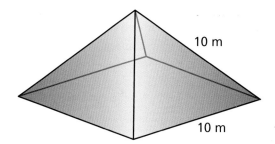

17 A door wedge is in the shape of a triangular prism with a right-angled triangle cross-section. A line is drawn from *D* to *F* for decoration. What is the length of line drawn?

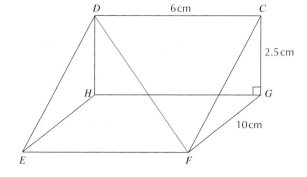

18 A climbing frame is in the shape of a triangular prism.

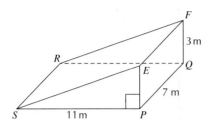

a Stuart is at *S* and Fay is at *F*. If Fay uses the edges of the frame, what is the shortest distance she needs to travel to reach Stuart?

b There is a zip wire rope directly between *F* and *S*. If Fay travels directly across the climbing frame using the zip wire (from *S* to *F*), what is the distance she has to travel?

c How much shorter is her journey using the zip wire than along the edges of the climbing frame?

19 Reina is having her central heating replaced.

The plumber wants to leave an 8 m pipe in her garage.

The garage is the shape of a cuboid and is 6 m long, 5 m wide and 3 m high.

Will the pipe fit in her garage? Give a reason for your answer.

★ **20** Sandy and Jo hire a van to move house.

The storage space in the van is the shape of a cuboid and is 4 m long, 2 m wide and 2.2 m high.

They roll up their living room carpet. The length of the roll is 5 m.

Will the carpet fit in the van? Give a reason for your answer.

21 A runner rates running up a hill in terms of the steepness of its gradient like this:

- 0–10%: easy
- 11–30%: manageable
- 31–75%: slightly challenging
- 76–100%: very challenging
- more than 100%: painful.

Using these ratings, describe the following hills, giving a reason for your answer.

a

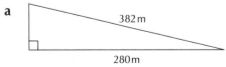

b

22 Building regulations require a roof to have a minimum gradient of 0.27.

This roof is an isosceles triangle. Does the roof meet the building regulation? Give a reason for your answer.

10 m

15 m

⊙ Activity

1 A spiral shape can be created using right-angled triangles.

Start with a right-angled triangle with base 2 cm and height 1 cm (this is the shaded triangle at the bottom of the diagram).

The second right-angled triangle has the hypotenuse of the first as its base and also has a height of 1 cm.

The shape grows by adding right-angled triangles onto each other.

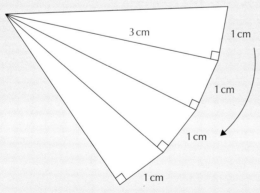

1 cm

1 cm

1 cm

1 cm

2 cm

a Draw this shape as far as you can (as accurately as possible).

b Measure the hypotenuse of the last triangle you have drawn.

c What is the percentage increase in the size of the hypotenuse between the first and last triangles drawn?

2 A second type of spiral shape can also be created using right-angled triangles.

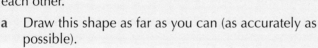
3 cm   1 cm

1 cm

1 cm

1 cm

Start with a right-angled triangle at the top with base 3 cm and height 1 cm.

The second right-angled triangle (below the first) has the base of the first as its hypotenuse and also has a height of 1 cm.

The shape grows by adding right-angled triangles under each other.

a Draw this shape as far as you can (as accurately as possible).

b Measure the hypotenuse of the last triangle you have drawn.

c What is the percentage decrease in the size of the hypotenuse between the first and last triangles drawn?

(continued)

3 There is a formula for quickly calculating the space diagonal of a cube or cuboid. The formula is:

space diagonal $= \sqrt{a^2 + b^2 + c^2}$

The formula could be used to calculate the space diagonal of the solid object in Exercise 20A Question 13 part **a**.

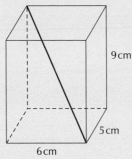

9 cm

5 cm

6 cm

Space diagonal $= \sqrt{6^2 + 5^2 + 9^2}$

$= \sqrt{36 + 25 + 81}$

$= \sqrt{142} = 11.9$ cm (3 s.f.)

Try using this method with Question 13 parts **b** and **c**, and with Questions 14, 19 and 20.

- I can use Pythagoras' theorem within a two-stage calculation. ★ Exercise 20A Q20

For further assessment opportunities, see the Case Studies for Unit 2 on pages 255–257.

Case studies

Nursing home

This case study focuses on the following skills:

- solving a problem involving time management **(Ch. 15)**
- solving problems using speed, distance and time **(Ch. 21)**
- investigating a situation involving a gradient **(Ch. 17)**.

A nursing home needs to build a ramp for wheelchair access into the building.
The local council's requirements are:

- maximum gradient of 1 : 20 for ramps and paths over 5 m in horizontal length
- maximum gradient of 1 : 15 for ramps and paths under 5 m in horizontal length.

This is their design:

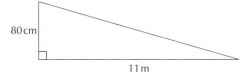

80 cm

11 m

The builder must be ready to *start* the job at 10.30 am at the nursing home.

He knows that it will take him 15 minutes to unload his van.

It usually takes him 20 minutes to pick up the material he needs from the depot.

The depot is 6 miles from the nursing home and he can drive between the depot and the nursing home at an average speed of 25 mph.

The depot is 12 miles from his house and he can drive from his house to the depot at an average speed of 40 mph.

1   Does the ramp meet the council's requirements? Give a reason for your answer.

[1S, 1P, 1C]

2   The builder needs to unload the van before he can start the job. What time does he need to leave home in order to be ready to start the job on time?   **[1S, 2P, 1C]**

Total 7 marks

Hidden treasure

This case study focuses on the following skills:

- construct a scale drawing, including choosing a scale **(Ch. 11)**
- recording measurements using a scale on an instrument **(Ch. 22)**.

An orienteering course has been designed by the local council for children to participate in a treasure hunt.

From the start, the first treasure is 200 metres away on a bearing of 125°.

From this first point the second treasure is 460 metres away on a bearing of 238°.

The third treasure is 50 metres due west of the starting point.

When participants have discovered the final treasure, they must return to the starting point.

To make it fair, each year the course must be no longer than (1200 ± 50) metres.

    a Using a suitable scale, make a scale drawing of the treasure hunt.   **[1S, 2P]**

    b What is the bearing and distance from the second treasure to the third treasure?   **[1C]**

    c Is the course within the required tolerance? Give a reason for your answer.?   **[1S, 1C]**

Total 6 marks

Community centre

> **This case study focuses on the following skills:**
>
> - investigating a situation involving a gradient **(Ch. 17)**
> - solving a problem involving a composite shape which includes part of a circle **(Ch. 18)**
> - using Pythagoras' theorem within a two-stage calculation **(Ch. 20)**.

A painter has been asked to redecorate a room in a community centre.

The painter has a ladder which is 3.5 metres long.

The Health and Safety Executive (HSE) states that a ladder must be placed against a wall with a maximum gradient of 4 and a minimum gradient of 2.

The diagram shows the dimensions of the room (not to scale).

The room has two windows: one is a rectangle and the other is a composite shape consisting of a rectangle and semi-circle.

It also has two doors: the main door into the room and a cupboard in the room. The doors are made of wood and are **not** painted.

A 5-litre tin of paint costs £39.98.

1 litre of paint covers 16 m².

The painter charges £120 for his labour.

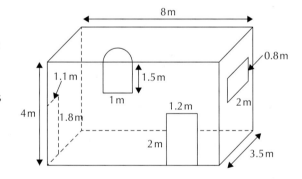

    1 The painter places the ladder 80 centimetres (horizontally) from the wall.

        a How far up the wall does the ladder reach?   **[1S, 1P]**

        b Using your answer to part **a**, find the gradient the ladder makes when placed against the wall.   **[1S]**

        c Does the ladder meet the requirements of the HSE? Give a reason for your answer.

    [1C]

        d The painter moves the ladder so that it is placed 1 metre from the wall. Does it meet the requirements of the HSE? Give a reason for your answer.   **[1P, 1C]**

6 marks

2 The community centre's budget is £350. Is this enough? Give a reason for your answer.

[1S, 3P]

4 marks

Total 10 marks

Drinks package

> **This case study focuses on the following skills:**
> - analysing a problem involving container packing **(Ch. 13)**
> - solving a problem involving the volume of a composite solid **(Ch. 19)**
> - explaining decisions based on the results of calculations **(Ch. 21)**.

A company sells cans of juice in the shape of a cylinder.

The can has a height of 12.2 cm and a volume of 330 ml.

Each empty can weighs 17 g and the juice weighs 1.13 g per ml.

The company needs to pack 24 cans into cardboard boxes for delivery to their clients.

They have two options of box packaging:

One layer of cans packed into a box containing 6 cans by 4 cans.

Two layers of cans packed into a box with 3 cans by 4 cans on each layer.

The box surrounds the cans exactly so that they are packed tightly.

1 **a** What is the diameter of each can of juice? [1S, 1P]

     b By calculating the surface area of each box, state which box uses packaging material more efficiently [1S, 2P, 1C]

6 marks

2 **a** How much do the **contents** of one box weigh? [1S, 1P]

     b The company decides that the contents of a box should have a maximum weight of 15 kg. What is the maximum number of cans that could now be packaged?

[1P, 1C]

4 marks

Total 10 marks

21

Selecting and using appropriate numerical notation and units and selecting and carrying out calculations

This chapter will show you how to:

- round answers to the nearest significant figure or 3 decimal places
- use appropriate methods to check the answers to calculations
- solve problems by adding, subtracting, multiplying and dividing whole numbers and decimals
- recognise and work with mixed fractions and improper fractions
- solve problems by calculating with percentages, decimal fractions and fractions
- solve problems that involve perimeter, area and volume
- solve problems that involve speed, distance and time
- solve problems that involve ratio, direct and inverse proportion
- interpret the results of calculations to make decisions
- explain decisions based on the results of calculations.

You should already know:

- how to select and use appropriate numerical notation
- how to select and use appropriate numerical units
- how to use the correct order of operations
- how to round answers to the nearest significant figure or 2 decimal places
- how to use appropriate methods to check calculations
- how to solve problems by adding, subtracting, multiplying and dividing with whole numbers
- how to solve problems by calculating with percentages, decimal fractions and fractions
- how to solve problems that involve perimeter, area and volume
- how to solve problems that involve time, rates, ratio and proportion
- how to interpret the results of calculations to make decisions
- how to explain decisions based on the results of calculations.

Numeracy

As you study the Numeracy unit you will further develop the following skills:

- selecting and using the appropriate numerical notation and units
- thinking about a problem and choosing the correct mathematical processes to solve the problem
- solving problems in context and relating the answer back to this context
- using measuring instruments and reading scales
- making and explaining decisions by interpreting the results of calculations.

Most of the questions in the exercises in this chapter use the contexts of number, money, time and measurement. You need to become familiar with the notation and units used in these contexts.

Solving problems

Many of the examples in this chapter demonstrate a particular approach to solving problems. When solving a problem, ask yourself these questions:

- *What bit of maths is this question asking for?* Identify the topic or concept, and then decide on a **strategy**.

- *What calculation do I need to do?* Show the **process** or stages of your working.

- *What does the numerical answer mean?* **Interpret** your answer, relating it to the context in the question.

- Sometimes you need to make a **decision** based on your interpretation. If so, you might need to give a reason for your decision.

Rounding answers

This section revises the rules for rounding numbers, how to round decimal numbers and how to round to the nearest significant figure.

When you round off an answer there are some guidelines to follow.

- If there are several steps in your calculation, wait until you get the final answer before you round off. If you round too early you could introduce an error into your final answer.

- Think carefully about the context of the question before you round. Make sure your answer makes sense.

- Check the question you are solving. It may tell you the accuracy you have to give your answer to.

- The usual steps for rounding are:

  o find the digit in the place-value column you need to round to, for example, the second decimal place

  o look at the next digit

  o if this next digit is 4 or less, do not change the digit

  o if this next digit is 5 or more, round the digit up

- You need to make sure that your answer makes sense. Sometimes, because of the context of the question, you will need to 'round up' even though the usual rule tells you to 'round down'.

Example 21.1

Nav is redesigning his garden. He is planning to lay a lawn measuring 4.5 m by 3.2 m.

Each turf rolls that he plans to buy will cover $0.75 \, \text{m}^2$. Each roll costs £7.

At the gardening centre, he finds that if he buys more than 20 rolls, the price is reduced to £6.50 per roll

How much will it cost Nav to buy the turf for his lawn?

The area of the lawn is $4.5 \times 3.2 = 14.4\,\text{m}^2$. ● ──┤ First work out the area of the lawn to be turfed. ├

Number of turf rolls is $14.4 \div 0.75 = 19.2$ rolls ● ──────┤ Each turf roll covers $0.75\,\text{m}^2$. ├

Nav needs to buy at least 20 rolls to complete his lawn. ● ──┤ Although 19.2 rounds down to 19, the context of the question requires 19.2 to be rounded up, as 19 rolls wouldn't be enough to cover the whole lawn. ├

If Nav buys 20 rolls, the turf will cost him $20 \times £7 = £140$.

If he buys 21 rolls, the turf will cost him $21 \times £6.50 = £136.50$.

Nav should buy 21 rolls so that he can take advantage of the special offer.

Rounding to 3 decimal places

When you round a number to 3 decimal places you look at the digit in the fourth decimal place.

Example 21.2

Round each of these numbers to 3 decimal places (3 d.p.).

a 7.9635   **b** 9.3102   **c** 14.9996

a $7.963\underline{5} \approx 7.964$ (3 d.p.) ● ──────┤ Remember: 5 or more rounds up. ├

b $9.310\underline{2} \approx 9.310$ (3 d.p.) ● ──┤ 4 or less rounds down. Keep the zero to show you have rounded to 3 d.p. ├

c $17.999\underline{6} \approx 18.000$ (3 d.p.)

What are significant figures and why are they used?

Science and technology are underpinned by mathematics. Using science we aim to understand how everything in (and beyond) the world we live in works. Scientific experimentation plays a large part in helping us in our quest and mathematics gives us the tools to analyse and describe the connections and patterns we discover. When we experiment we take measurements and these measurements always have a limited level of accuracy. These measurements are often used in calculations which give results that also have a limited level of accuracy. Depending on the type (or number) of calculations we do, we find our final answer can be far less accurate than the original measured value. **Significant figures** (sometimes called significant digits, and abbreviated to sig figs or s.f.) are used to keep track of the accuracy of measurements and allow us to be confident in the accuracy of calculations involving measured values.

Rules when using significant figures

Errors are a large part of experimental science and how we handle them in calculations is very interesting but can get quite complicated. At this level, you don't need to know how to do this but if you are interested you could ask your teacher (maths or science) to explain a little more. For now, all you need to know are the rules that are used to decide if a digit is significant or not. Follow these rules and you won't go wrong.

Rule 1: Non-zero digits are always significant

946.27 has 5 significant figures. Note that the '9' is the first (or most) significant figure as it has the largest place value and represents 900. The '7' is the least (or fifth) significant figure as it has the smallest place value and represents 7 hundredths.

Rule 2: One (or more) zeros with a non-zero digit on either side of it is significant

30050.303 has 8 significant figures.

Rule 3: Leading zeros are never significant

The leading zeros in the number **0.0000**45 are shaded. Only the 4 and the 5 are significant. This number has 2 significant figures.

Rule 4: Trailing zeros may or may not be significant

670000 has only 2 significant figures. The trailing zeros in a whole number are not significant unless we know for certain that the number is accurate. For example, usually the number 100 has 1 significant figure. However, in the context of a statement such as 'there are 100 pennies in one pound', then, since we know this to be exact, we say that (in this context), 100 has 3 significant figures.

Trailing zeroes are usually not significant for whole numbers but are significant in decimal numbers. So, **3.4600** has 5 significant figures. Usually we wouldn't include the zeros when writing a number such as 3.4600. Writing the zeros indicates that this is a measured value and that we are certain that the number is accurate. An example of this would be a set of electronic scales that are accurate to 0.001 of a gram. If you weigh an item on these scales and get a reading of 2.400 you can be sure that the last two digits have been measured accurately and this number has 4 significant figures.

Example 21.3

Round 0.054 36 to 2 significant figures.

$\rightarrow$ 0.054 (to 2 s.f.)

Use **Rule 3**: leading zeros are never significant.

Look at the 3rd significant figure; 3 is less than 5 so round down. Note that 0.05436 rounded to 2 decimal places gives 0.05.

Example 21.4

Round 2.3685 to 3 significant figures.

2.36_85_

$\rightarrow$ 2.37 (to 3 s.f.)

Note that 2.3685 rounded to 3 decimal places gives 2.369.

Example 21.5

Round 5300.045 500 to 7 significant figures.

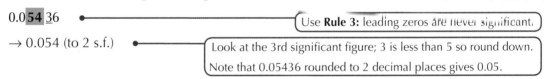

$\rightarrow$ 5300.046 (to 7 s.f.)

Use **Rule 2** (zeros with a non-zero digit on either side are significant), so the zeros in the tens and units columns are significant here. 5300.045 500 is accurate to 6 decimal places.

Example 21.6

Round 70.499 to 4 significant figures.

70.49_9_

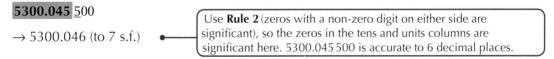

$\rightarrow$ 70.50 (to 4 s.f.)

If asked for 4 significant figures you must give 4, even if it means including zeros.

Example 21.7

Round 5349 to 1 significant figure.

5349

→ 5000 (to 1 s.f.) ●————————————(Include zeros to preserve place value.)

Exercise 21A

★ **1** Round these numbers to 3 decimal places.

  a 2.3524 **b** 23.5609 **c** 3.4504 **d** 345.9487 **e** 0.0095

  f 13.4183 **g** 2.0409 **h** 5.1697 **i** 0.03995 **j** 17.9997

★ **2** Round each of the following numbers to 2 significant figures.

  a 7890 **b** 3.461 **c** 287 **d** 0.476 **e** 23.569

  f 3489 **g** 0.0054 **h** 6500 **i** 0.000457 **j** 230005

3 Kevin the baker uses the following ingredients to make blueberry muffins.

Estimate the weight of one muffin to 3 significant figures.

Ingredients

Makes 120 muffins

- 1.1 kg butter
- 2.5 kg plain flour
- 2.5 kg caster sugar
- 20 eggs (0.37 kg)
- 1.25 litres milk
- 20 teaspoons baking powder
- 10 teaspoon salt
- 2.25 kg fresh blueberries

4 The table shows the total length of public roads in different countries around the world and the population of each country.

  a Round the length of all public roads in each country to 2 significant figures.

  b Round the population of each country to 2 significant figures.

  c Use your answers to parts **a** and **b** to find the number of metres of road per person.

  d Which three countries have the greatest length of road per person? Why do you think these countries have more road than the others?

Country	Length of all public roads (km)	Population
UK	398350	64181775
USA	6430351	316156000
Canada	1408800	35141542
Germany	644480	80399300
Scotland	55420	5254800
Ireland	98752	4385400
Australia	810624	23671313
India	4163000	1210569573

Using appropriate checking methods for calculations

When you calculate an answer, it is important to know if your answer makes sense.

It would not, for example, make sense for a shop assistant to earn £35000 per week.

It would, however, make sense for the shop assistant to earn £350 per week.

You should always do a 'rough' calculation to see if your answer fits with the given context of the question.

One way to estimate an expected answer is to round each number to 1 significant figure and then calculate mentally.

Example 21.8

Work out the answer to each calculation. First, estimate the answer by rounding each number to 1 significant figure.

a 21×53 **b** $3132 \div 62$ **c** 3.9×12.32 **d** $18 \div 4.9$

a $21 \times 53 \rightarrow 20 \times 50 = 1000$

The answer is approximately 1000.

The exact answer is 1113.

b $3132 \div 62 \rightarrow 3000 \div 60 = 50$

The answer is approximately 50.

The exact answer is 50.516...

c $3.9 \times 12.32 \rightarrow 4 \times 10 = 40$

The answer is roughly 40.

The exact answer is 48.048.

d $18 \div 4.9 \rightarrow 20 \div 5 = 4$

The answer is roughly 4.

The exact answer is 3.673...

You can also use your knowledge of number facts to help check for errors.

For example, when a number is multiplied by 5 the last digit of the answer is always a 0 or a 5.

So you know that the last digit in the answer to 346×5 will be 0 or 5.

In fact $346 \times 5 = 1730$, which ties in with what you know.

There are other rules for multiplication facts that you can find out for yourself.

You can also use inverse properties to check your solutions. By working backwards you can find out if your answer is correct.

Example 21.9

Check if the answers to these calculations are correct.

a $420 \div 7 = 60$ **b** $856 - 89 = 765$

a $420 \div 7 = 60$

$60 \times 7 = 6 \times 10 \times 7$ •——— Multiplication is the inverse ('opposite') of division, so the answer can be checked by calculating.

$= 42 \times 10 = 420$

So $420 \div 7 = 60$ ✓ •——— The calculation is correct.

b $856 - 89 = 765$ ——— Addition is the inverse of subtraction, so the answer can be checked by doing the calculation $765 + 89$. If the initial calculation is correct, the answer will be 856.

$765 + 89 = 854$ •———

So $856 - 89 = 765$ ✗ •——— The calculation is wrong.

Note that in part **b**, it was not necessary to complete the addition to see that the initial calculation was wrong. The unit digits of the numbers to be added are 5 and 9 and $5 + 9 = 14$, so the last digit of the answer to $765 + 89$ is a 4 and not a 6.

Exercise 21B

★ **1** Without actually finding the answer, explain how you know that each of these calculations is wrong.

    a $15 \times 32 = 482$ **b** $346 \div 56 = 9.18$ **c** $7689 + 247 = 9936$

    d $670 - 46 = 626$ **e** $13.25 + 18.98 \div 10 = 3.223$ **f** $146 \div 3 = 50$

★ **2** By rounding each number to 1 significant figure, estimate the answer to each of these calculations. Do **not** use a calculator.

    a $11489 - 3257$ **b** $\dfrac{20.5 + 56.23}{12.65 - 2.67}$

    c 56.4×78.97 **d** $4972 \div 24.99$

3 When Dougie shops in a supermarket he tries to estimate the total bill before he goes through the checkout. What do you think was his estimated total for these items in his basket?

    bananas 78p   loaf 70p   margarine £1.80   apples £1.45

    soup £1.80   rolls 90p   magazine £4.99   milk 67p

4 Louise drove from New York to Los Angeles, a distance of 2775.4 miles. Her overall average speed was 43.2 mph. Louise drove for 11 hours each day.

    She left New York at 1130 am on Tuesday 6 January 2015.

    Estimate the day and time that she will arrive in Los Angeles.

5   Jimmy uses oil for his central heating. His storage tank measures 189 cm by 176 cm by 126 cm. The gauge on the tank shows that the tank is 33% full. The oil that he needs costs 48.56 pence per litre.

Estimate the cost for Jimmy to fill the tank with oil.

6   A group of 25 Girl Guides are going to stay in a youth hostel for 4 nights.

They plan to eat baked beans on toast for their breakfast each morning.
Each portion of baked beans is 120 g.

A large cylindrical tin of baked beans has a radius of 7.3 cm and a height of 18.2 cm.

By estimating the volume of baked beans in one tin, how many tins of beans should they buy?

> ⚠ 1 cm³ of baked beans weighs 1 g.
> The volume of a cylinder is $V = \pi r^2 h$.

7   Estimate the value each arrow is pointing to.

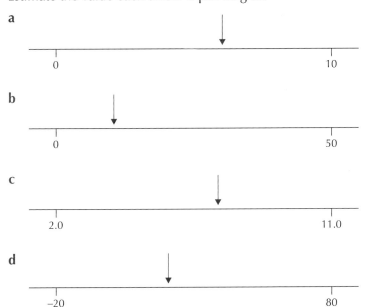

a   0 ———————— 10

b   0 ———————— 50

c   2.0 ———————— 11.0

d   −20 ———————— 80

● Activity

Research the 'rules of divisibility'.

You can use the rules of divisibility when you are dividing whole numbers to see if the number you are dividing by is a factor.

Use the rules that you have found to find all the factors that are less than 10 for each of these numbers.

    a 345   b 984   c 3333   d 45 009   e 237 632

Solving problems by adding, subtracting, multiplying and dividing with whole numbers and decimals

When solving problems you need to be able to:

•   choose an appropriate strategy

•   explain or write down the process that you have used

★ **5** Priorsland High School are holding an Enterprise Fair in the school hall. Students at the school can hire a stall for different events.

- They can sell goods that they have produced. These students pay for their stall, but keep the money they raise for themselves. All proceeds from the stall hire go to charity.

- They can run a tombola/lucky dip/guess the lucky square stall. These students do not need to pay the fee to hire the stall.

All proceeds from the stall hire go to charity.

There are 35 stalls available for hire at £7.50 each.

21 groups of students hire stalls to sell their own goods.

The rest of the stalls are hired by students raising money for charity.

The charity stalls raise, on average, £32.47 each.

What is the total amount that the school raises to give to charity?

6 Poppy's Party Planners organises birthday parties for children. The table shows the prices they charge.

Item	Cost
Venue hire	£65
Entertainment	£55
Food/drink (per person)	£3.75
Cups/plates/napkins (per person)	£0.99
Table cloths (1 for every 10 people)	£2.50
Small birthday cake (for up to 10 people)	£27
Large birthday cake (up to 20 people)	£38
Party bags (per person)	£5

VAT of 20% must be added to these prices.

Sam and Gerard ask Poppy's Party Planners to plan a party for their children Gwen and Eden along with 14 of their friends. How much will the party cost Sam and Gerard in total?

7 A local cash and carry warehouse does a stocktake of the sweets they have available. The table shows the results of the stocktake.

Name of sweet	Number of 1 kg bags	Buying price per bag
Gummy Penguins	5	£9.56
Space Riders	7	£6.72
Gob Stoppers	2	£8.67
Allsports	8	£7.97
Fruit Gumdums	5	£4.65
Bazoo Jims	12	£8.43

The selling price of each sweet is calculated by adding 30% to its buying price.

How much will the cash and carry warehouse get in total if they sell all these sweets?

Solving problems by calculating with fractions, decimal fractions and percentages

Fractions, decimals and percentages are three different ways of finding part of an amount.

In this section you will look at how the three are connected as well as use them to solve problems from finance, business and science.

Some of this work should be familiar to you from N4.

This section extends the work that you have already covered to more difficult problems.

Finding percentages and fractions of amounts and shapes

Example 21.11

A publisher offers a discount of 13% on orders over £550.

Stewart's High School maths department orders copies of a new textbook costing £13.50 per copy.

What is the least number of copies of the textbook that they must order to qualify for the discount?

How much will they pay if they order this number?

£550 ÷ £13.50 = 40.740 740...

> To find the least number that they should order, divide the value that they must exceed by the price of one textbook.

They will need to order 41 books to qualify.

> Round up.

Total price = 41 × £13.50 = £553.50

> Work out total price of order before discount.

Discounted price will be 87% of £553.50

= 87 ÷ 100 × £553.50

> A discount of 13% means that they will pay (100 − 13)% = 87% of the total price before the discount.

= £481.55

Example 21.12

a Calculate the length of minor arc AB.

b Calculate the area of major sector AOB.

Give both answers to 3 significant figures.

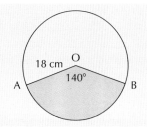

a The minor arc is a fraction of the circumference of the circle based on the size of the angle at the centre of the minor (shaded) sector.

$$\text{Arc length} = \frac{\theta}{360} \times \pi d$$

> Use the formula for arc length. See Chapter 18 for more on using the formula.

$$\text{Length of AB} = \frac{140}{360} \times \pi \times 36$$
$$= 43.982\,297\ldots\text{cm}$$
$$= 44.0\,\text{cm (3 s.f.)}$$

> The diameter is twice the radius. Remember that to find a fraction of an amount, you divide by the denominator (the bottom number) and then multiply by the numerator (the top number).

b The major sector is part of the area of the circle based on the size of the angle at the centre of the unshaded sector.

Area of a sector = $\dfrac{\theta}{360} \times \pi r^2$

> Use the formula for sector area. See Chapter 18 for more on using the formula.

Area of AOB = $\dfrac{220}{360} \times \pi \times 18^2$

> Angle of major sector is $360° - 140° = 220°$.

$= 622.035\,345\ldots \text{cm}^2$

$= 622\,\text{cm}^2$ (3 s.f.)

Percentage increase and decrease

Example 21.13

Mr Jones invests £600 in a savings account that pays 5% per annum interest.

If he leaves the interest in the account each year, how much will he have in total after 3 years?

There are two methods for answering this question.

Method 1

Capital at start of year 1	£600
Interest for year 1	5% of £600 = 0.05 × £600 = £30
Capital at end of year 1	£600 + £30 = £630
Capital at start of year 2	£630
Interest for year 2	5% of £630 = 0.05 × £630 = £31.50
Capital at end of year 2	£630 + £31.50 = £661.50
Capital at start of year 3	£661.50
Interest for year 3	5% of £661.50 = 0.05 × £661.50 = £33.08 (to nearest 1p)
Capital at end of year 3	£661.50 + £33.08 = £694.58

> 5% = 5 ÷ 100 = 0.05

Method 2

By the end of year 1, the capital will have increased by 5%, so Mr Jones will have 100% + 5% = 105% of the amount he started with.

By the end of year 1, he will have £600 × 1.05

> 105% = 105 ÷ 100 = 1.05

By the end of year 2, he will have amount at end of year 1 × 1.05 = £600 × 1.05 × 1.05

By the end of year 3, he will have amount at end of year 2 × 1.05 = £600 × 1.05 × 1.05 × 1.05

Total at end of year 3 = £600 × 1.05^3

$= £694.58$ (to the nearest 1p)

Both methods give the same answer, but method 2 is quicker, especially if you are calculating the interest for more than 3 years.

Example 21.14

Jenny bought a new house for £195 000 in January 2007.

The value of the house rose by 5.5% per year for the next 3 years.

The value of the house then decreased by 4% for the next 2 years.

Jenny decided to sell her house in January 2012.

What price could she expect to get for her house?

Give your answer to 3 significant figures.

The value of the house after 5 years needs to be calculated in two parts, using the percentage increase and decrease given.

The value after 3 years will be
£195 000 × 1.055³ = £228 977.07

For the first part, use a multiplier of
100% + 5.5% = 105.5% = 1.055

The value after the next 2 years will be
£228 977.07 × 0.96² = £211 025.27

The multiplier for the following
2 years will be 100% − 4% = 96% = 0.96

The value of the house is now
£211 000 (3 s.f.), so Jenny could expect to get about £211 000 for her house.

Example 21.15

A 1-year-old car is worth 80% of its purchase price.

A 1-year-old Fiat Panda is worth £12 240.

How much was the purchase price of the Fiat Panda?

80% of original = £12 240

The value of the 1-year-old car is 80% of the original price.

1% of original = £12 240 ÷ 80 = £153

100% of original = £153 × 100 = £15 300

So the purchase price of the Fiat Panda was £15 300.

Equivalence between fractions, decimal fractions and percentages

It is easy to convert between fractions, decimals and percentages.

- To change a fraction to a decimal, divide the numerator by the denominator.
- To change a fraction or a decimal into a percentage, multiply it by 100.
- To change a decimal to a fraction, make the denominator 10, 100 or 1000 depending on the number of decimal places, then simplify if possible.
- To change a percentage to a fraction, make the denominator 100, then simplify if possible.
- To change a percentage to a decimal, divide it by 100.

It is often easier to use one form rather than another, depending on the information given and the context of the question.

To compare fractions, decimals and percentages, they need to be changed into the same format first.

Example 21.16

Arrange these in order from smallest to largest.

$\frac{2}{3}$, 70%, $\frac{3}{5}$, 0.65, 66%, 0.706

It is easier to compare percentages or decimals, so change all of these into either decimals or percentages.

$\frac{2}{3}$ = 2 ÷ 3 = 0.667 (3 d.p.) = 66.7% (1 d.p.)

70% = 0.7

$\frac{3}{5}$ = 3 ÷ 5 = 0.6 = 60%

0.65 = 65%

66% = 0.66

0.706 = 70.6%

From smallest to largest:

    as percentages: 60%, 65%, 66%, 66.7%, 70%, 70.6%

    as decimals: 0.6, 0.65, 0.66, 0.667, 0.7, 0.706

$\frac{3}{5}$, 0.65, 66%, $\frac{2}{3}$, 70%, 0.706   ●————⟨ Write the amounts in their original formats in order. ⟩

Unless a format is specified, you can use whichever format – fraction, decimal or percentage – is most suitable for a given question.

If you don't have a calculator, it is often easiest to use a fraction rather than a percentage or a decimal.

If you do have a calculator, it is often easiest to change a percentage into a decimal before typing it in.

Example 21.17

Bernard buys an antique teddy bear for £320.

He sells it for £368 six months later.

Calculate Bernard's percentage profit.

The percentage profit is $\dfrac{\text{actual profit}}{\text{buying price}} \times 100\%$ ●————⟨ Actual profit is £368 − £320 = £48. ⟩

$= \dfrac{48}{320} \times 100\%$

$= 15\%$

Example 21.18

A shop is giving a 17.5% discount on all its goods.

What is the sale price of television originally priced at £850?

Do **not** use a calculator.

17.5% = 10% + 5% + 2.5%

10% of £850 is £850 ÷ 10 = £85

5% of £850 is £85 ÷ 2 = £42.50

2.5% of £850 is £42.50 ÷ 2 = £21.25

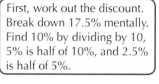

First, work out the discount. Break down 17.5% mentally. Find 10% by dividing by 10, 5% is half of 10%, and 2.5% is half of 5%.

17.5% of £850 is £85 + £42.50 + £21.25 = £148.75.

So the sale price of the television will be £850 − £148.75 = £701.25.

Exercise 21D

★ 1 Copy and complete the following table to show the fraction, decimal and percentage equivalents.

Fraction	Decimal	Percentage
$\frac{1}{2}$		
	0.25	
		10%
	0.75	
$\frac{2}{3}$		
		80%
	0.$\dot{3}$	
$\frac{3}{4}$		
		20%
	1.4	
		250%

2 Which of the following is less than $\frac{2}{5}$?

    a 0.35   **b** 39%   **c** 2.5   **d** $\frac{9}{20}$ **e** 0.41

3 Harry is shopping for a new suit. He finds the same suit in two different stores.

- Clark's Outfitters are offering $\frac{1}{3}$ off the recommended retail price of £199.
- South's Clothes Shop is giving a 40% discount on their normal price of £220.

Which shop should he buy the suit from? Use your working to explain your answer.

4 In January 2005 the number of children attending nursery in Moray was 1389. The number increased by 2.5% for each of the next 3 years.
How many children attended nursery in 2008?

5 The number of school teachers in Scotland this year is 52188. This is $7\frac{1}{2}$% more than last year.

Calculate how many teachers there were in Scotland last year.

6 Rohini is buying a second-hand car. The cash price of the car is £6300. The garage is offering a hire purchase deal of a deposit of $33\frac{1}{3}$% of the cash price followed by 24 payments of £185.
Rohini decides to buy the car using the hire purchase deal. Calculate how much extra this will cost her. Give your answer as a percentage of the cash price.

7 420 people are surveyed about their shopping habits. They had to choose from the following four categories:

- I never shop online
- I shop online and in a real store
- I only shop online
- I never shop.

The survey's results showed that:

- 55% said they shopped online and in a real store
- $\frac{1}{5}$ said they never shopped online
- 3% said they never shop.

Find the total number of people who responded for each of the four categories.

★ 8 Catriona and Gill are shopping for a new three-piece lounge suite.
They find two different offers online for the same furniture.

Which offer should they choose? Use you working to justify your answer.

Homeland Furniture
Three-piece suite
Deposit £300
24 months interest-free credit at £45
per month
£200 final payment

Fancy stuff online
Three-piece suite £1800
6% off if you buy
now!

9 Helen is going to invest £4500 in a long-term savings account.

She can afford to leave her money and any interest that it earns for 4 years.

She finds the following information on the internet.

Bank	Savings rate p.a. (%)
Bank of Scotia	1.2
Hebrides Savings Bank	0.79
Harold's Bank	2.4
Zebra Saving	3.47

> p.a. (per annum) means 'per year'.

a Assuming that the interest is compound, calculate how much money she will have in her account after 4 years with each of these banks.

b How much more money will she have after 4 years if she invests with the bank offering the highest interest rate than the bank with the lowest rate?

Solving problems that involve volume, area and perimeter

In this section you will solve problems that involve perimeter and area of compound shapes.

> See Chapters 18 and 19 for more on area and volume.

You will also solve problems involving the volumes of solid objects including cubes, cuboids, triangular prisms, cylinders spheres and hemispheres.

Example 21.19

A soft drinks manufacturer is designing a cylindrical can that must contain at least 330 ml of the drink. The can must be no more than 6 cm in diameter.

a Calculate the minimum height that the can will need to be for this diameter.

b What is the area of metal needed to manufacture one of these cans?

Give each answer to 3 significant figures.

Use these formulae:

Volume of a cylinder is $V = \pi r^2 h$

Surface area of a cylinder is $A = \pi dh + 2\pi r^2$

a $V = \pi r^2 h$

$330 = \pi \times 3 \times 3 \times h$ ← *Substitute the values given into the formula. Remember that radius = diameter ÷ 2.*

$330 = 9\pi \times h$ ← *Rearrange the equation to make h the subject and solve.*

$h = \dfrac{330}{9\pi}$

$= 11.671362\ldots$

$= 11.7\,\text{cm (3 s.f.)}$

The can needs to be at least 11.7 cm in height.

b $A = \pi dh + 2\pi r^2$

$A = \pi \times 6 \times 11.7 + 2 \times \pi \times 3 \times 3$

$= 220.539\,804\ldots + 56.548\,668\ldots$

$= 277.088\,472$

$= 277\,\text{cm}^2$ (3 s.f.)

The area of metal needed for one can is $277\,\text{cm}^2$.

Exercise 21E

1 Calculate the perimeter and area of this shape.

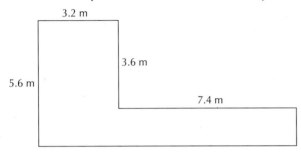

2 Find the volume of the following solids. Where appropriate give your answer to 3 significant figures.

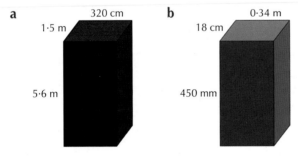

★ **3** Mary is going to tile the floor and walls of her new wet room. She has chosen tiles that measure 200 mm by 200 mm.

The plan of the floor of Mary's wet room is shown below.

The wet room is 2.6 m high.

The wet room has a door that measures 120 cm by 210 cm in the middle of the 1.5 m wall.

There is one window measuring 150 cm by 80 cm in the middle of the 2.4 m wall.

She is not going to tile the ceiling.

a Estimate the number of tiles that Mary will need to buy. Show your calculations to justify your estimate.

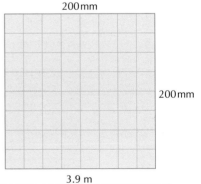

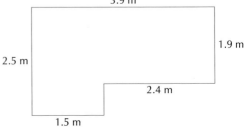

The tiles are only available in packs of 10. Each pack costs £42.50.

She will also need to buy tile adhesive and grout.

The tile adhesive costs £39.99 for each 20 kg tub. Each tub will cover 6.8 m².

The grout costs £18.99 per pack and contains enough to grout 18 m² of completed tiling.

b Calculate how much it will cost Mary for the materials to tile her wet room.

The tiler she employs to lay the tiles charges £26 per hour. The tiler estimates that they will take 30 hours to complete the job. VAT of 20% must be added to the cost of the materials and labour.

c Calculate the total price for Mary to tile her wet room.

★ **4** A soft drinks manufacturer is creating a new cylindrical container for its new fruit drink. Each can must hold 250 ml of the drink.

Market research has shown that consumers want

* the diameter of the container to be between 5 cm and 8 cm

* the height of the container to be between 7 cm and 10 cm.

The machinery that makes the containers operates to 5 mm intervals.

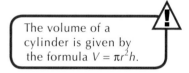
The volume of a cylinder is given by the formula $V = \pi r^2 h$.

What dimensions could the manufacturer use for the new container?

5 A rugby pitch is marked up as shown in the scale diagram below. The pitch is the entire green area.

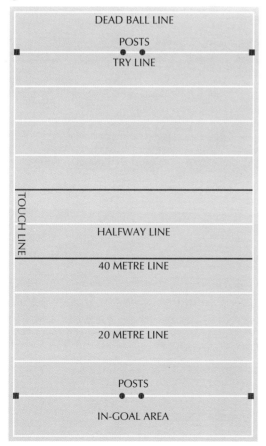

Scale: 1 cm = 10 m

    a What is the perimeter of the pitch?

    b What is the total length of all the white lines?

    c Calculate the area of the pitch behind the posts.

    d Calculate the total area of the pitch.

6 A decorative edging for a garden lawn is produced by cutting a sector from a cylindrical piece of wood as shown in the diagram. The sector is used as the edging.

Each piece of edging is 75 cm long.

The diameter of the cylinder is 8 cm.

Calculate the volume of wood in each section of the edging.

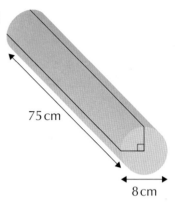

7 A wooden door with a decorative window is shown in the diagram.

The door measures 0.9 m by 2.1 m. The window is a sector of a circle with radius 28 cm. The angle at the centre of the sector is 130°. The glass in the window is surrounded by a lead edging strip.

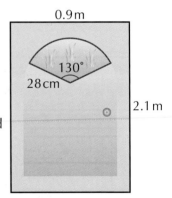

    a Calculate the length of lead edging strip that will be needed to make one of these decorative windows. Give your answer to 3 significant figures.

    b Calculate the area of the wooden part of the door.

    Give your answer to 3 significant figures.

> ⚠ The length of a sector arc is given by the formula arc length $= \dfrac{\theta}{360} \times \pi d$.
> The area of a sector is given by the formula $A = \dfrac{\theta}{360} \times \pi r^2$.

8 A tent is in the shape of a triangular prism.

The end of the tent is an isosceles triangle with width 140 cm and a sloping side of 190 cm. The tent is 2.6 m long.

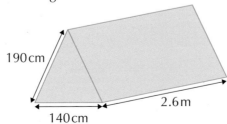

The manufacturer claims the tent has a volume of 3 cubic metres. Is this claim justified? Show your calculations.

9 Jenny has started a small craft business to supplement her regular wage. She is making candles for a local school fair. Each candle is in the shape of a cylinder with a cone on top.

The cylinder has a diameter of 7 cm and a height of 9 cm.

The overall height of the candle (not including the wick) is 14 cm.

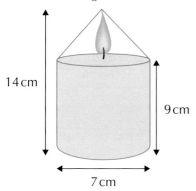

Jenny has said she will provide 50 candles for the fair. She has a 20 litre tub of wax in stock to make the candles. Will she be able to fulfil her order? Justify your answer.

> ⚠️ The volume of a cylinder is given by the formula $V = \pi r^2 h$.
>
> The volume of a cone is given by the formula $V = \frac{1}{3}\pi r^2 h$.

10 A glass paperweight is in the shape of a hemisphere. The diameter of the paperweight is 6 cm.

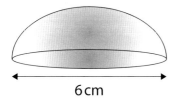

Calculate the volume of glass in the paperweight.

> ⚠️ The volume of a sphere is given by the formula $V = \frac{4}{3}\pi r^3$.

★ **11** A wash hand basin made of granite is in the shape of a cuboid with a hemisphere cut from it.

The cuboid has dimensions 45 cm by 35 cm by 18 cm.

The hemisphere has a diameter of 19 cm.

a Calculate the volume of granite, in cubic centimetres, in one of these wash hand basins.

The packaging that the sink is delivered in weighs 3.75 kg. The delivery company states that for items over 70 kg two people must lift the package.

b Calculate the total weight of the sink and packaging and justify whether one or two delivery people are required to lift it.

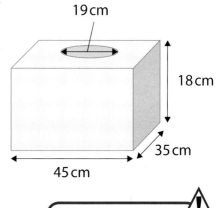

> ⚠️ The volume of a sphere is given by the formula $V = 4/3\pi r^3$. 1 cm³ of granite weighs 2.75 g.

Solving problems that involve time, ratio, proportion and rates

Rates and time

Speed is an example of a **rate**.

It is the distance travelled per unit of time.

There are three related formulae for calculating speed, distance and time.

Use this triangle to remember them:

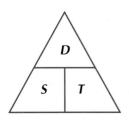

$$S = \frac{D}{T} \qquad D = S \times T \qquad T = \frac{D}{S}$$

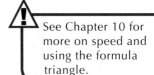
See Chapter 10 for more on speed and using the formula triangle.

Example 21.20

Craig drives 180 miles from Elgin to Glasgow for a meeting that starts at 10:40.

He expects to travel at an average speed of 50 mph.

He needs to arrive 30 minutes before the meeting to allow time to park his car and walk to the venue.

At what time should he leave Elgin if he wants to arrive on time at the meeting?

Craig will need to calculate how long his journey will take him and then 'work back' from the time that he needs to arrive.

$$T = \frac{D}{S}$$

Use the formula to work out how long it will take to drive from Elgin to Glasgow.

$$= \frac{180}{50}$$

$$= 3.6 \text{ hours}$$

3.6 hours = 3 h 36 min

0.6 hour is 0.6 × 60 minutes = 36 minutes.

Total journey time = 3 h 36 min + 30 min

$$= 4 \text{ h } 6 \text{ min}$$

Departure time is 10:40 − 4 h 6 min

Subtract the total journey time from the time of the meeting to work out the departure time.

$$= 06:34$$

Ratio and proportion

Ratio and **proportion** are used to describe the relationship between two or more values or quantities.

When solving problems involving proportion, always think about the context of the question to decide if you are expecting to get a bigger answer or a smaller answer than the number in the question.

Example 21.21

Jenni and her three friends decide to go to the cinema.

The tickets cost them £37 in total.

At the last minute, six other friends decide to join the cinema trip.

How much will all the tickets cost now?

There are two methods of answering this question.

Method 1

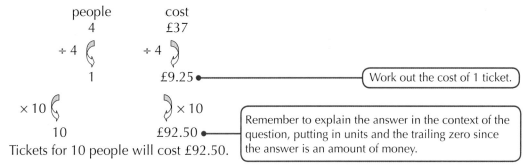

Tickets for 10 people will cost £92.50.

Method 2

people   cost
4   £37   [Compare the number of people in the group with the total price for the tickets.]
10   x

$x = \dfrac{10}{4} \times £37$ [Multiply the cost of 4 tickets by ratio of the new group size to the old group size.]

$= £92.50$ [Check that you have multiplied by the correct ratio: the cost of 10 tickets will be more than the cost of 4 tickets.]

Tickets for 10 people will cost £92.50.

Example 21.22

An aid organisation provided emergency supplies to a refugee camp.

They delivered enough food and water to feed the 2500 people in the camp for 5 days.

An extra 1500 people arrived at the camp the day after the food and water was delivered.

How many days would the food and water now last?

In this question, we are comparing the number of people in the camp with the length of time that the food and water would last.

$2500 \times 5 = 12\,500$ portions   [Work out how many individual daily portions of food and water were delivered.]

On the first day, 2500 of these portions were used, leaving 10 000 portions unused.

On day 2 there were 4000 people in the camp.

$10\,000 \div 4000 = 2.5$ [Work out how long the food and water will last for the new number of refugees. Divide the number of portions by the number of people.]

The food would last another 2.5 days.

Example 21.23

Ollie, Molly and Polly win £3780 in the lottery.

They split the money in the ratio $3:4:2$ respectively.

How much will each person get?

$3 + 4 + 2 = 9$ parts ●————(Work out the total number of parts in which the money will be split.)

Ollie will get 3 parts out of the 9 parts.

Ollie will get $\frac{3}{9} \times 3780 = 3780 \div 9 \times 3 = £1260$ ●————(Work out how much each person will get.)

Molly will get $\frac{4}{9} \times 3780 = 3780 \div 9 \times 4 = £1680$

Polly will get $\frac{2}{9} \times 3780 = 3780 \div 9 \times 2 = £840$

Check: $1260 + 1680 + 840 = 3780$ ✓ ●————(Check that the amounts add to the original total.)

So Ollie will get £1260, Molly will get £1680 and Polly will get £840.

Exercise 21F

1   The recipe for pancakes uses milk, flour and water in the ratio $9:4:1$.

    The milk and water are measured in millilitres and the flour is measured in grams.

    If 6300 ml of milk is used for one batch of pancakes, how much flour and water is needed?

2   Alex and Bert are hiking together. They know that their average walking speed is 5 km/h. They stop for lunch after they have walked for 13.5 km.

    Calculate how long they walked for before they stopped for lunch. Give your answer in hours and minutes.

3   Jenny, Helen and Andy are counting the number of steps they take at work as part of a health-at-work day. They use pedometers to count their steps.

    Jenny takes 1512 steps in 7 hours, Helen takes 1628 steps in 7 hours and 24 minutes and Andy takes 1510 steps in 7 hours and 33 minutes.

    Who takes the most steps per hour?

★   4   Seonaid goes on holiday to France for one week and then to Switzerland for two weeks. She changes money to euros (€) and Swiss francs (CHF) during her holiday. The table shows the rates of exchange that were used when Seonaid changed her money.

| £1 = €1.27 |
| £1 = 1.53CHF |
| €1 = 1.21CHF |

    a   She changes £800 into euros before she leaves for France. How many euros does she receive?

    Each day that she is in France, she spends an average of €45.

    b   When she goes to Switzerland, she changes her remaining euros to Swiss francs. The bank she uses will only change multiples of €10. How many Swiss francs will she get?

    Each day that she is in Switzerland, she spends an average of 50CHF.

c   When she returns home, she changes her remaining Swiss francs back to pounds sterling. Again the bank will only change multiples of 10CHF. How many pounds sterling will she get for her remaining Swiss francs? Give your answer in pounds and pence.

5   These are the ingredients needed to make 12 treacle scones:

250 g flour
pinch salt
1 teaspoon cinnamon
25 g butter
25 g caster sugar
1 rounded tablespoon black treacle
120 ml milk

Lucy is baking treacle scones to sell to raise money for charity. She makes 40 scones in total.

a   How much of each ingredient does she need?

The ingredients for the 40 scones cost £2.37 in total. She calculates that the electricity she uses costs £1.31.

b   If Lucy sells the scones for 35p each, how much money will she raise for charity after she has deducted her expenses?

★ 6   The table shows the records for athletic events of different distances. Calculate the average speed for each event.

Event	Time
100 m	9.58 s
200 m	19.19 s
400 m	42.18 s
800 m	1 min 40 s
1500 m	3 min 26 s
10 km	26 min 44 s
30 km	1 h 27 min 38 s

7   A plane has a maximum range of 1350 km when flying at an average speed of 600 km/h.

a   If the plane takes off from Aberdeen at 09:30, what is the latest time that the plane could land?

b   Use a map or the internet to research all the possible destinations this plane can reach without refuelling.

8   Four friends compare their texting speed. Use the information below to work out who can text the fastest.

Jake: 45 characters in 30 s   Kelly: 76 characters in 1 min

Yvonne: 23 characters in 10 s   Pete: 65 characters in 45 s

How much should the group leader charge each pupil so that all the costs are covered?

3   Each pupil is carrying approximately 15 kg of luggage and each adult is carrying approximately 25 kg.

   a   Calculate the total weight of the group's luggage.

   b   What is the mean weight of the luggage taken on the trip?

4   The risk assessment estimates that 15% of the group will suffer a minor injury. How many of the group are expected to have a minor injury?

5   The table shows the average maximum daily temperature and number of days of rainfall in April in Oban for the last 10 years.

Temperature (°C)	4.5	11.0	6.2	7.6	5.5	13.1	16.3	9.9	8.6	8.9
Number of days with rainfall	6	17	7	9	4	24	25	19	16	17

   a   What is the range of temperatures in the table?

   b   Calculate the mean and standard deviation for the temperatures in April in Oban. Give your answers to 2 significant figures.

   c   The mean and standard deviation for the temperatures in Oban in June are 17°C and 2.5°C respectively. Make two valid comments comparing the temperatures in Oban in April and in June.

   d   What is the probability that April will have fewer than 10 days of rain?

6   As part of the outdoor education award, all of the pupils must complete the walk shown in red in the map below.

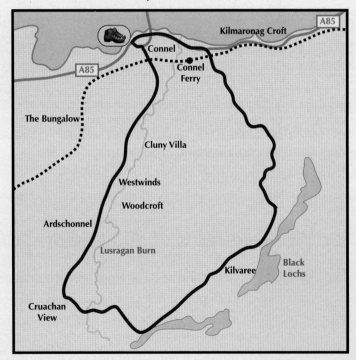

   a   If the scale of the map is 1 : 25 000, estimate the total length of the route.

   b   How long will it take the group to walk the route at an average speed of 2 km/h?

Key questions

- I can round answers to the nearest significant figure or 3 decimal places. ★ Exercise 21A Q1, Q2

- I can use appropriate methods to estimate the answers to calculations. ★ Exercise 21B Q1, Q2

- I can solve problems by adding, subtracting, multiplying and dividing whole numbers and decimals. ★ Exercise 21C Q5

- I can recognise and work with mixed fractions and improper fractions. ★ Exercise 21D Q1

- I can solve problems by calculating with percentages, decimal fractions and fractions. ★ Exercise 21D Q8

- I can solve problems that involve perimeter, area and volume. ★ Exercise 21E Q3, Q4, Q11

- I can solve problems that involve speed, distance and time. ★ Exercise 21F Q6

- I can solve problems that involve ratio, direct and inverse proportion. ★ Exercise 21F Q4

- I can explain decisions based on the results of calculations. ★ Exercise 21F Q10, 12

- I can interpret the results of calculations to make decisions. ★ Activity pages 285–286.

For further assessment opportunities, see the Case Studies for Unit 2 on pages 322–326.

22 Recording measurements using a scale on an instrument to the nearest marked, minor unnumbered division

This chapter will show you how to:

- read scales on measurement instruments to the nearest marked, unnumbered division
- record measurements using different scales and units
- convert between different units of measurement
- interpret and explain decisions based on the results of measurements.

You should already know:

- the metric units of length, weight, capacity and temperature
- how to convert between metric measurements
- the importance of making an estimate before reading a measurement
- how to use measuring instruments with straightforward scales
- how to record measurements accurately given different scales.

Reading scales

Reading scales is an important skill in everyday life.

Measuring length, volume, weight, temperature and angles all involve being able to read a scale from a measuring instrument such as a ruler, protractor, weighing scales or a thermometer.

The scale on most measuring instruments will have major and minor divisions.

The **major divisions** are marked with an amount, but the **minor divisions** often are not marked with amounts.

The first thing you need to do when using a scale is to work out the size of each minor division of the scale.

Example 22.1

The scale shown can measure up to 200 g in weight.

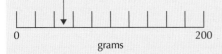

0
grams
200

At what weight is the arrow pointing?

Each division is worth 200 g ÷ 10 = 20 g.

0 20 40 60   200
grams

> There are 10 divisions shown on the scale between 0 g and 200 g.

> The arrow is almost exactly midway between 40 g and 60 g.

The arrow is pointing at 50 g.

Example 22.2

The scale shown can measure between 50 m and 70 m in length.

50   70
metres

Copy the scale and mark it to show 63 m.

Each division is worth 20 m ÷ 5 = 4 m.

> There are five divisions between 50 m and 70 m.

63 m

50   54   58   62   66   70
metres

> 63 m will be 1/4 of the way between 62 m and 66 m.

Exercise 22A

★ **1** Estimate the number that each arrow is pointing to on the number lines below.

a

0   i   ii   10

b

0   iii   iv 20

c

0   v   vi   30

d

0   vii   viii   200

2 Write down the values indicated by **a**, **b**, **c** and **d** on these weighing scales.

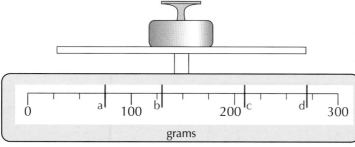

0   a   100   b   200 c   d   300
grams

3 How much, in litres, do these six jugs hold altogether?

i 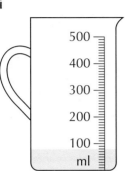 ii 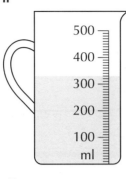 iii

iv 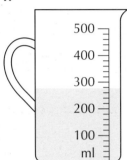 v 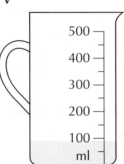 vi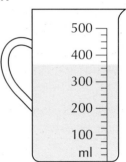

4 What is the temperature shown on each thermometer?

a

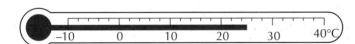

b

c

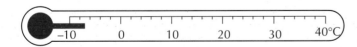

d

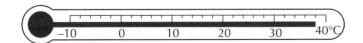

5   Temperature is usually measured in either degrees Fahrenheit, °F, or in degrees Celsius, °C.
    The thermometers below show the relationship between the two scales.

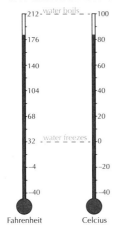

The relationship between degrees Celsius and degrees Fahrenheit is given by the formula

$$°C = \frac{5}{9}(°F - 32)$$

The thermometers below each show a reading in °F. Use the formula to convert each
temperature into °C.

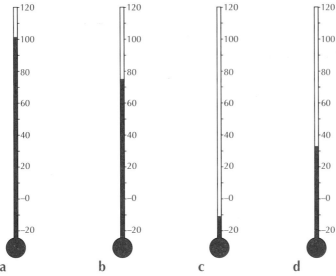

    a   b   c   d

6   The dial shows the speed at which a space rocket is travelling.

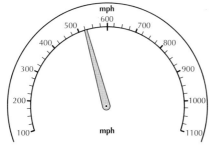

To escape the earth's gravity it will need to reach a speed of 820 mph.

How much faster will the rocket need to travel to achieve this speed?

Interpreting and recording on scales

There are many situations in everyday life where you will need to interpret a scale or record a measurement on a scale.

Exercise 22B

1   Recipe books often specify cooking temperature in degrees Celsius. If the oven that you are using has a scale in degrees Fahrenheit, or is a fan oven or is fuelled by gas, you need to work out the equivalent temperature so that you cook the dish at the correct temperature.

The table below shows the conversions needed for different types of oven.

	Electricity (°F)	Electricity (°C)	Electricity, fan (°C)	Gas mark
Very cool	230	110	90	$\frac{1}{4}$
	250	120	100	$\frac{1}{2}$
Cool	280	140	120	1
	300	150	130	2
Moderate	320	160	140	3
	360	180	160	4
Moderately hot	375	190	170	5
	390	200	180	6
Hot	430	220	200	7
	445	230	210	8
Very hot	465	240	220	9

Convert the temperatures given in each part, then sketch this temperature on a cooker dial similar to the examples shown here.

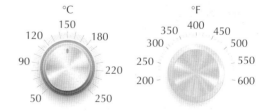

a 320°F to °C (no fan)   b gas mark 4 to °C (fan)   c 200°C to °F (no fan)
d gas mark 6 to °C (fan)   e gas mark 1 to °F   f 170°C (fan) to °F

★ 2   Dry ingredients in baking might be given in grams or in ounces. The usual conversion used between the two in baking is:

1 ounce = 25 grams

Convert each of these amounts. Copy the weighing scale shown for each and show the converted amount.

⚠ This is not an accurate conversion. Recipes are designed to take this inaccuracy into account. You should not weigh some ingredients in ounces and others in grams in the same recipe or the recipe might not work.

a 175 g to ounces

b 14 ounces to grams

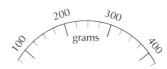

3 When baking, different amounts of liquid are required for different recipes. Match these amounts to the amounts labelled i, ii, iii and iv.

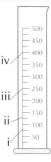

a 130 ml **b** 0.4 litre **c** $\frac{1}{4}$ litre **d** 0.08 litre

4 The diagram shows part of a scale.

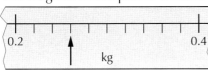

Grace says that the arrow is pointing at 0.23 kg. Explain why she is wrong and draw a similar scale to show the correct position of 0.23 kg.

5 Paddy is driving from Edinburgh to Hull, a distance of 263 miles. The fuel gauge in his car is shown.

1 gallon = 5.546 litres

The tank in Paddy's car can hold 80 litres of fuel. The car can travel an average of 23 miles per gallon of fuel.

Will Paddy have enough fuel to get to Hull without refilling the tank? Use your working to justify your answer.

6 François, who is from France, is on holiday in Scotland. The speedometer on his car is shown.

The section of road that he is travelling on has a speed limit of 50 mph.

Exercise 22C

1 For each list of measurements or quantities, find the odd one out. Justify your choice.

    a   456 cm   4.56 m   4560 mm   0.456 km

    b   7800 g   7.8 kg   7800 m*l*   7 kg 800 g

    c   2000 m   0.2 km   200 000 cm   2000 000 mm

    d   13 *l*   1300 m*l*   130 c*l*   1.3 *l*

2 Oliver is using kitchen scales to weigh out pasta. The recipe requires 0.65 kg of pasta. Explain how he could use the scales shown to weigh this amount of pasta.

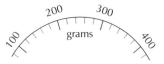

★ 3 Cherie is planning to do some DIY. She wants to buy three lengths of wood, each measuring 1.5 m by 2 cm by 3 cm.

At the local store she found lengths of wood described as:

    a   150 mm × 20 mm × 300 mm

    b   1500 mm × 2 mm × 30 mm

    c   150 mm × 20 mm × 3 mm

    d   1500 mm × 20 mm × 30 mm

Which of these are suitable for Cherie to buy for her DIY?

★⚙4 Kevin is going on holiday. He is allowed to take a suitcase weighing up to 20 kg. The scales below show the results when Kevin stands on the bathroom scales first on his own and then holding his suitcase.

Kevin realises that he has forgotten a number of items. What is the maximum weight that these extra items could be? You must explain your answer.

5 This scale shows the amount of rain that fell on the town of Bettyhill during one day in April.

The mean annual rainfall for Bettyhill is 1.83 m.

Compare the amount of rain measured on this day with the annual mean.

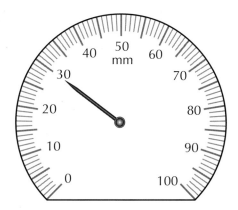

2

6 The speedometer shows speeds in mph and km/h.

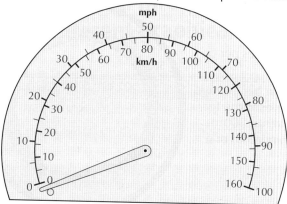

a Use the speedometer to change 50 mph into km/h.

b Use the speedometer to change 120 km/h into mph.

c Graham drives 60 km in 40 minutes. Hannah drives at 55 mph.

Who is driving faster? You must explain your answer.

★ 7 A normal human body temperature is
(36.8 ± 0.4)°C.

Louise's temperature is shown on the thermometer in °F.

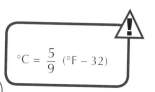

$$°C = \frac{5}{9}\ (°F - 32)$$

Is Louise's temperature within the normal limits?

8 Amy is organising a birthday party for 20 children. She plans to give each child 125 ml of juice. She knows from previous experience that about one third of the children will ask for a second glass of juice. The jug shows how much juice Amy has made.

How much more juice will she need to make to cater for the party? Give your answer in millilitres and show your working.

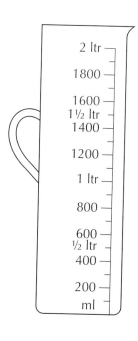

9

★ 1

6 The boxplots show how long a group of students took to do a fitness activity before and after some months of training.

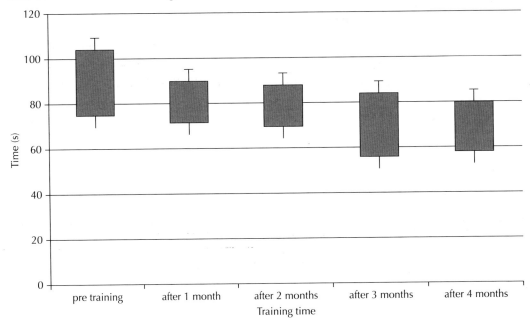

Comment on how the group of students progresses over the 4-month period.

7 The graphs show the performance of the share price of three companies over a 6-month period.

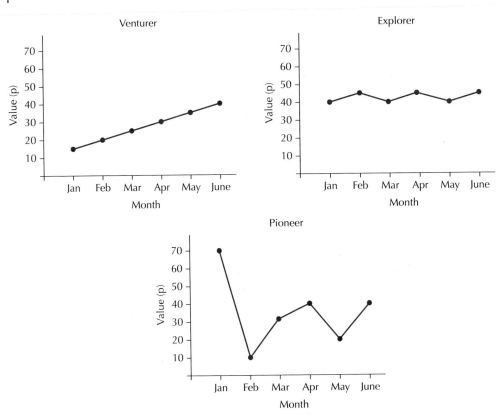

For

Veronica wants to invest money in shares. Use the graphs to decide which company would be:

a   the safest option for her investment

b   the option that may give her the highest return for her money.

8   It can cost several thousand pounds to buy a wind turbine but most of the costs are due to construction and buying energy conversion systems, such as inverters and batteries. Once installed, however, there are no fuel costs and you only need to pay for maintenance.

The scatter diagram below shows the installation costs of different models of wind turbines and the expected cost of the energy (in kilowatt hours, kWh) each will produce in an area with an average wind speed of 4 m/s.

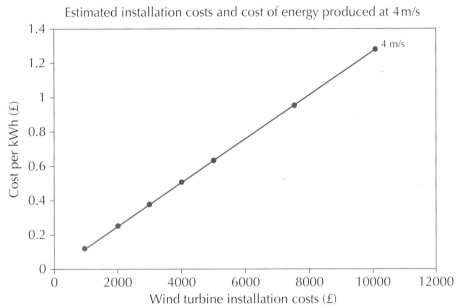

Estimated installation costs and cost of energy produced at 4 m/s

a   How much would a turbine that produces energy at 80p per kWh cost to install?

b   The table below shows the installation costs and cost of energy in an area that has an average speed of 8 m/s. Copy the scatter diagram above and add the data for 8 m/s.

Installation cost (£)	1000	2000	3000	4000	5000	8000	10000
Cost of energy per kWh (£)	0.01	0.08	0.1	0.15	0.2	0.3	0.4

c   Draw a best-fitting line for the 8 m/s data.

d   Compare the two areas with average wind speeds of 4 m/s and 8 m/s. In which area would it be best to site a wind turbine? Justify your answer.

(GO!) Activity

When they retire, Ian and Rhona are going to buy a villa in Spain which they will rent out to holidaymakers. They are considering six villas in different locations.

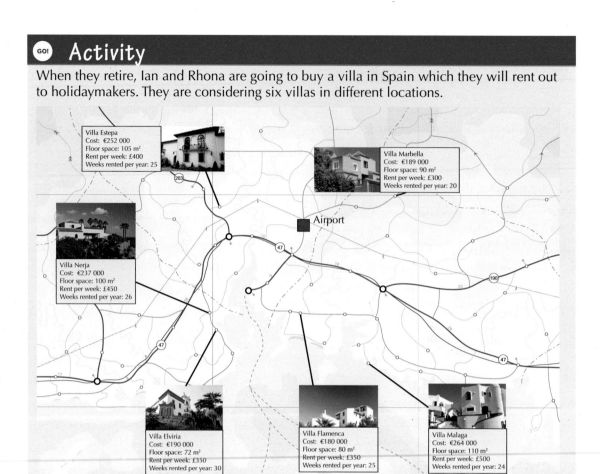

Villa Estepa
Cost: €252 000
Floor space: 105 m²
Rent per week: £400
Weeks rented per year: 25

Villa Marbella
Cost: €189 000
Floor space: 90 m²
Rent per week: £300
Weeks rented per year: 20

Villa Nerja
Cost: €237 000
Floor space: 100 m²
Rent per week: £450
Weeks rented per year: 26

Airport

Villa Elviria
Cost: €190 000
Floor space: 72 m²
Rent per week: £350
Weeks rented per year: 30

Villa Flamenca
Cost: €180 000
Floor space: 80 m²
Rent per week: £350
Weeks rented per year: 25

Villa Malaga
Cost: €264 000
Floor space: 110 m²
Rent per week: £500
Weeks rented per year: 24

£1 = €1.2

By comparing the information for the six villas, give Ian and Rhona advice to help them decide which option would be best in the long term.

- I can extract information from different graphical forms and interpret data. ★ Exercise 23A Q2, Q5

- I can make and justify decisions based on the interpretation of data. ★ Exercise 23B Q1, Q3

For further assessment opportunities, see the Case Studies for Unit 2 on pages 322–326.

24 Making and justifying decisions based on probability

This chapter will show you how to:

- recognise **patterns**, **trends** and **relationships** and use these to state the probability of an event happening
- use evidence from the **interpretation of probability** to justify decisions
- analyse the probability of **combined events**
- identify the effects of **bias**
- describe probability through the use of percentages, decimal fractions and ratio to **make** and **justify decisions**.

You should already know:

- how to calculate simple probability
- how to use probability to make predictions
- how to add and multiply fractions
- how to convert between fractions, decimal fractions and percentages.

Trends, patterns and relationships

The **trend** of a graph is the description of the **relationship** between the two quantities plotted on the graph.

The graph could be a line graph or a scatter graph.

The trend is the overall picture rather than the description of each individual part of the graph.

The trend in a graph can be thought of as a **pattern** that connects the different variables.

The pattern is often a change in one variable over a period of time. An example of this would be a graph showing the changes in temperature over a 12-month period.

- The pattern of the temperatures would be that it was warmer in the summer and cooler in the winter over the 12-month period.

- The trend in temperature change would be seen by comparing the same month over several years, rather than different months within a single year.

Example 24.1

This graph shows the number of deaths on British roads between 1967 and 2003 for all road users and for pedestrians.

Describe the trends shown in the graph.

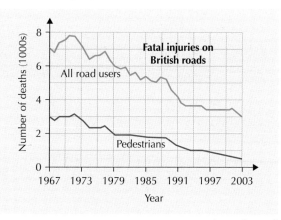

(continued)

Both lines in the graph come down from a higher number of deaths in 1967 to a lower number of deaths in 2003.

There was a slight increase in deaths between 1967 and 1973, but the overall trend for each line is downwards, meaning that there was a decrease in road deaths for all road users and also for pedestrians over the period shown in the graph.

Using evidence from the interpretation of probability to justify decisions

Probability is an assessment of the likelihood of a particular outcome. It can be calculated using evidence from past events, such as weather or survey data, or from known likelihoods, such as rolls of dice or card games. This probability can then be used to forecast future outcomes.

Probability is used in many different fields such as geography, insurance, sports, gambling, horse-racing and the finance industry.

Example 24.2

A local authority is going to build a new school in one of two towns, Aberton and Blairkenny. The pie charts represent the age profiles of the two towns.

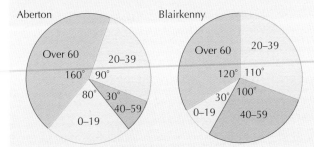

Aberton has 75 40–59 year olds.
Blairkenny has 500 40–59 year olds.

a **i** Calculate how many people are in each age group in each town.

   ii Calculate the total population of each town.

b Find the probability that a person chosen at random from the two towns:

   i lives in Aberton and is aged 0–19

   ii lives in Blairkenny and is aged 0–19

   iii lives in Blairkenny and is aged 20– 39

   iv lives in either Aberton or Blairkenny and is aged 40–59.

c What percentage of each town's population is in each age group?

d Using the information from the pie charts, decide whether the school should be built in Aberton or in Blairkenny. You must explain your answer.

e What other information might be useful to the local authority before it finally decides where the school will be built?

a i In Aberton, there are 75 people in the 40–59 age group. This sector of the pie chart is measured as 30°.

> Use proportion to work out how many people are represented by 1°.

number of people   angle (°)

$\begin{array}{cc} 75 & 30 \\ \div 30 \downarrow & \downarrow \div 30 \\ 2.5 & 1 \end{array}$

So 1° represents 2.5 people.

> Use this information to work out how many people are in each age group.

Other age groups in Aberton:

0–19: $80 \times 2.5 = 200$

20–39: $90 \times 2.5 = 225$

Over 59: $160 \times 2.5 = 400$

In Blairkenny, there are 500 people aged 40–59.

$500 \div 100 = 5$

> As before, use proportion to work out how many people are represented by 1°.

So 1° represents 5 people.

Other age groups in Blairkenny:

0–19: $30 \times 5 = 150$

20–39: $110 \times 5 = 550$

Over 59: $120 \times 5 = 600$

ii The total population of Aberton is $75 + 200 + 225 + 400 = 900$

The total population of Blairkenny is $500 + 150 + 550 + 600 = 1800$

b i The probability that a person lives in Aberton and is aged 0–19 is

$$\frac{200}{900 + 1800} = \frac{200}{2700} = \frac{2}{27}$$

> Add the total populations of the two towns to work out the denominator.

ii The probability that a person lives in Blairkenny and is aged 0–19 is

$$\frac{150}{2700} = \frac{5}{90} = \frac{1}{18}$$

iii The probability that a person lives in Blairkenny and is aged 20–39 is

$$\frac{550}{2700} = \frac{55}{270} = \frac{11}{54}$$

iv The probability that a person lives in either Aberton or Blairkenny and is aged 40–59 is

$$\frac{75 + 500}{900 + 1800} = \frac{575}{2700} = \frac{23}{108}$$

(continued)

    iii Aberdeen

    iv Lerwick

The table shows the average maximum daily December temperature in the same seven locations.

A television company is planning to do a live broadcast on Christmas day from one of these seven locations. They need to have a minimum temperature of 4°C and the highest probability of snow.

Location	Average temperature
London	6°C
Birmingham	5°C
Aberporth	5°C
Glasgow	4°C
Aberdeen	4°C
Belfast	5°C
Lerwick	2°C

b From which location should they broadcast? You must explain your answer.

c Using a map of the UK to help you, what do you notice about the trend in:

    i the probability of snow at Christmas

    ii average temperatures in December?

★ **3** The Parents Association at Prestonfield High School are trying to decide the best time to hold a school Summer Fun Day.

They would like the day to be dry and the temperature to be at least 12°C.

The tables show the rainfall and temperature data for April, May and June between 2003 and 2012.

Number of days with rain

Month	2003	2004	2005	2006	2007	2008	2009	2010	2011	2012
April	9	18	15	16	8	7	20	10	14	9
May	20	11	17	18	17	15	8	21	9	20
June	13	18	13	10	13	10	19	13	7	19

Average minimum temperature (°C)

Month	2003	2004	2005	2006	2007	2008	2009	2010	2011	2012
April	9.6	5.4	10.5	8.9	9.5	12.4	9.6	11.9	14.5	12.8
May	12.3	7.4	11.2	4.4	9.5	14.9	12.3	13.9	11.2	14.1
June	16.2	9.4	15.9	16.3	15.9	16.3	15.7	17.6	16.4	13.9

a **i** Calculate the mean number of days with rain for each of April, May and June, to the nearest whole number, as a prediction of the number of days of rainfall in 2015.

    ii Use the mean numbers of days with rain and the number of days in the month to find the probability that it will rain on a given day in each of the three months. Give your answers as fractions.

    iii Express the probability of rain each month as a percentage.

    iii Based on this information, which month do you think would be best for the Summer Fun Day?

b **i** Using the data in the table on temperature, estimate the probability that the temperature will be at least 12°C in each of April, May and June.

ii What is the trend of temperature between April and June each year?

iii Using the information on minimum temperature, which month do you recommend for the Summer Fun Day?

c Use your answers to **a** and **b** to decide when the Summer Fun Day should be held.

d Gather weather information for your own area and suggest the best possible weekend for a local gala.

★ **4** In autumn 2007 a survey was carried out of commuting patterns in Great Britain. It identified several different modes of transport.

The table shows the percentages of people using different modes of transport. The commuting patterns in Greater London are different from those of the rest of Great Britain, so the results are given separately.

Main mode	Area of workplace		
	Greater London (%)	Rest of Great Britain (%)	Great Britain (%)
Car and van	37	76	71
Motorbike, moped, scooter	1	1	1
Bicycle	3	2	3
Bus and coach	14	7	7
National Rail	19	2	4
Underground	16	-	2
Walk	9	11	11
Other modes	1	1	1
Number of people (millions)	**3.34**	**21.48**	**24.83**

a What is the probability that:

i a commuter in London, chosen at random, travels to work on the Underground?

ii a commuter in the rest of Great Britain, chosen at random, travels to work by bicycle?

iii a commuter anywhere in Great Britain, chosen at random, travels to work by car?

Give each answer as a fraction in its simplest form.

Sometimes it is easier to understand the information if it is shown in a chart or graph.

b Draw two pie charts representing the proportion of people using each mode of transport in:

i London

ii the rest of Great Britain.

c Calculate the number of people using each mode of transport to commute to work and add this to your pie chart.

d If you were in charge of the budget for transport in Great Britain, what would your priorities in spending be in:

i London

ii the rest of Great Britain?

Comment on similarities and differences in priorities.

⚙ 5   The British Heart Foundation (BHF) has made a recommendation of a minimum amount of physical activity per week in order to lower the risk of coronary heart disease.

The BHF surveyed men and women aged 16 upwards in 2008. The table shows the percentage of men and women in different age categories who met the recommendations:

Age range (years)	16–24	25–34	35–44	45–54	55–64	65–74	75+
Men	53	49	44	41	32	20	9
Women	35	36	34	32	28	17	6

a   Illustrate the data for men and women on the same line graph.

b   Describe the trends shown in the graph for men and for women.

c   **i**   Why do you think that the trends follow this pattern?

    ii   Why do you think that the percentage of women in each age range is lower than that of men?

A town council has offered to provide funding for one new sport or activity group to be set up in the local community. They have four options:

- a dancersize class for women of all ages

- a football club for men under 35

- a walking group which offers walks during the day, mainly for the over-55s

- a training and fitness club run by the armed forces for anyone under 25.

d   Using the data above, which of the four options would you recommend to the council?

Give a reason for your answer.

Bias

When probability is significantly different from the probability expected in a fair trial, then the trial is **biased**. This could be as a result of an unfair instrument such as a weighted coin, dice or spinner, or it could be because the sample used was not a fair representation of the population. For example, the sample could include too many people from one age group, too many men or too many people from one ethnic background.

Example 24.3

A coin is tossed 100 times.

The results are 73 heads and 27 tails.

Explain how you know that this coin is biased.

If the coin is fair then **P(head)** = **P(tail)** = $\frac{1}{2}$

So the expected number of heads = expected number of tails = $\frac{1}{2} \times 100 = 50$

73 and 23 are very different from the expected result of 50 (even allowing for experimental variation), so it seems that the coin is biased. A result of 73 heads out of 100 suggests the coin is weighted towards the head.

Exercise 24B

1 A bio-engineering firm is testing a new fertiliser for potato crops. The tables show the results of tests using two different formulae for the new fertiliser.

Formula A

Crop weight per plant (to nearest kg)	Frequency
0	37
1	0
2	1
3	5
4	23
5	29
6	24

Formula B

Crop weight per plant (to nearest kg)	Frequency
0	3
1	18
2	24
3	39
4	58
5	4/
6	23

a Calculate the probability of achieving a crop weight of 4 kg using formula A.

b Calculate the chance of achieving a crop weight of 3 kg or more using formula B.

c The firm claims that formula A has a better than 93% chance of producing 4 or more kg of potatoes. How did they calculate this result and do you agree with this claim? Justify your answer.

d Which formula has the more reliable results? Justify your answer.

★ 2 The fitness of the S4 year group in Brecon Academy was assessed. To save time, a sample of S4 pupils was selected instead of testing the whole year group.

It was decided to record the length of time each pupil took to row 1000 m on a rowing machine.

The S4 National 5 PE class was used as the sample of 30 pupils.

The table shows the results.

Length of time taken, t (min)	Number of pupils
$5 \leq t$	2
$4 \leq t < 5$	6
$3 \leq t < 4$	16
$2 \leq t < 3$	4
$t < 2$	2

a Calculate the probability that a pupil took:

   i more than 5 minutes

   ii between 2 and 3 minutes

   iii less than 2 minutes.

   Give each answer as a fraction in its lowest terms.

b If there are 260 pupils in the year group, use these results to estimate the number of pupils who would take:

   i more than 5 minutes

   ii between 2 and 3 minutes

   iii less than 2 minutes.

   A second PE teacher decided to carry out the challenge for all the pupils in S4. The results are shown in the table.

Length of time taken, t (min)	Number of pupils
$5 \leq t$	27
$4 \leq t < 5$	90
$3 \leq t < 4$	120
$2 \leq t < 3$	12
$t < 2$	11

c Compare your answers in part **b** to the table above. Why do the results of the sample not reflect the results of the whole year group?

d Suggest how a sample could have been selected so that it better represented the entire S4.

3 Prior to the 1936 American presidential election, *The Literary Digest* carried out a poll to predict the results. Nearly 2.4 million people took part in the survey, which was one of the largest and most expensive polls ever conducted at that time. The table shows the results.

Name of candidate	Party	Number of votes in survey
Franklin D. Roosevelt	Democrat	1 031 862
Alf Landon	Republican	1 367 967

a Based on the survey, what is the probability, as a percentage, that a person selected at random would vote for:

   i Franklin D. Roosevelt

   ii Alf Landon?

The actual results in the 1936 election are recorded in the table below.

Name of candidate	Party	Number of votes in election
Franklin D. Roosevelt	Democrat	27 752 648
Alf Landon	Republican	16 681 862
William Lenke	Union	892 378
Norman Thomas	Socialist	187 910
Earl Browder	Communist	79 315
Other		53 586

b   Based on these results, what is the probability, as a percentage, that a person selected at random voted for:

   i   Franklin D. Roosevelt

   ii   Alf Landon?

c   Do you think the poll carried out by *The Literary Digest* used a fair sample? Give a reason for your answer.

★ 4   Shortly before the 2015 UK General Election a polling organisation carried out two opinion polls to predict the result of the election.

In the first opinion poll, 175 first year students studying politics at Edinburgh University were surveyed.

In the second opinion poll, 1137 shoppers on Princes Street, Edinburgh, were surveyed.

The voting intentions of each sample are shown in the table below.

Party	Students	Shoppers
Conservative	73	106
Liberal	21	75
Labour	40	315
SNP	35	583
Other/Not known	6	58
Total	175	1137

a   Calculate the percentage of each sample who stated that they were going to vote for each of the listed parties.

b   Why are the results different between the two samples?

c   Which opinion poll do you think would better represent the electorate as a whole?

d   The actual General Election result was not exactly the same as either opinion poll. Why do you think this was the case?

- I can recognise patterns, trends and relationships and use these to state the probability of an event happening. ★ Exercise 24A Q2

- I can use evidence from the interpretation of probability to justify decisions. ★ Exercise 24A Q3, Q4

- I can identify the effects of bias. ★ Exercise 24B Q2

- I can describe probability through the use of percentages, decimal fractions and ratio to make and justify decisions. ★ Exercise 24B Q4

For further assessment opportunities, see the Case Studies for Unit 2 on pages 322–326.

Case studies

Buying a car

> **This case study focuses on the following skills:**
>
> - extracting information from graphs and tables **(Ch. 23)**
> - making and justifying decisions based on probability **(Ch. 24)**.

When you buy a car for the first time there are a number of costs that you will have to consider. These include the initial price of the car and the annual running costs. Running costs include insurance, fuel, road tax and maintenance.

The bar chart shows the initial price of buying a range of second-hand cars and their annual running costs.

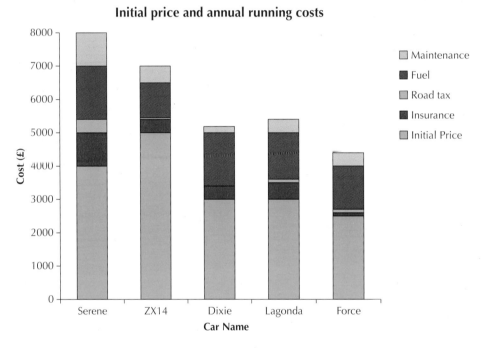

Initial price and annual running costs

Legend: Maintenance, Fuel, Road tax, Insurance, Initial Price

The table shows the number of each car on the road in 2014 and the number of reported breakdowns.

Car name	Number of cars on the road in 2014	Number of reported breakdowns
Serene	3098	450
ZX14	11670	1297
Dixie	24755	1600
Lagonda	2791	148
Force	16289	1534

1 a Which two cars have the same initial price? **[1C]**

b Estimate the cost of insuring the Lagonda. **[1S]**

c What is the difference in the annual fuel cost between the most fuel efficient car and the least? **[1P]**

d What is the percentage of the fuel cost compared to the total cost of buying and running the Dixie for one year? **[1P]**

4 marks

2 A salesperson for the Dixie states that the probability of it breaking down is less than 10%. Is this a valid claim? Use your working to justify your answer. **[1P, 1C]**

2 marks

Total 6 marks

Allotments

This case study focuses on the following skills:
- recognising and working with mixed fractions and improper fractions **(Ch. 21)**
- solving problems that involve perimeter and area **(Ch 18)**.

Janice and Michael both have allotments as they live near a city centre.

This is Janice's allotment: This is Michael's allotment:

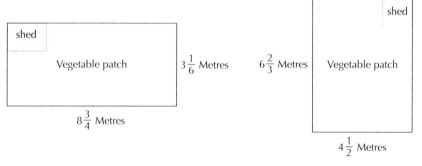

Each person has a shed in the corner of their allotment.

The ratio of the area of Janice's shed to vegetable patch is $1:6$.

The ratio of the area of Michael's shed to vegetable patch is $2:11$.

1 Whose allotment is bigger? Give a reason for your answer. **[1S, 1P, 1C]**

2 Whose vegetable patch is bigger and by how much? Give your answer correct to 3 significant figures. **[1S, 1P, 1C]**

Total 6 marks

Buying a sofa

This case study focuses on the following skills:

- investigating the impact of interest rates **(Ch. 5)**
- solving problems by calculating with percentages, decimal fractions and fractions **(Ch. 21)**
- recording measurements using a scale on an instrument **(Ch. 22)**.

Anji and Ben want to buy a new sofa that costs £998. The furniture shop offers two finance deals as an alternative to paying the full price immediately.

Deal 1	Deal 2
£100 deposit	No deposit
24 instalments of £41.95	36 payments of £31.25

The furniture shop's delivery charges are based on the weight of the item. The table shows the charges.

Weight	Delivery charge
Less than 50 kg	£20
50 kg up to 85 kg	£35
85 kg up to 110 kg	£45
110 kg up to 145 kg	£55
More than 145 kg	£65

The diagram shows part of a scale recording the weight of the sofa Anji and Ben bought.

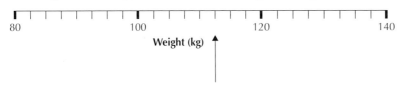

The value of the sofa will depreciate as soon as it is delivered. On average, a new sofa will lose 55% of its initial value in the first year, followed by 15% every year after that.

1. a How much extra does each deal cost compared with paying the full price immediately? **[2P]**
   b What is the total percentage interest paid on each deal? **[1P]**
   c Which is the better finance deal? Give a reason for your answer. **[1C]**

   4 marks

2. a What is the weight of the sofa? **[1S]**
   b How much will it cost for the sofa to be delivered? **[1C]**

   2 marks

3   a   How much do Ben and Anji expect the sofa to be worth in 4 years time?   **[1P]**

    b   How many years will it take for the value of the sofa to fall below £200?

[1P, 1C]

3 marks

Total 9 marks

Forth Road Bridge

This case study focuses on the following skills:

*   making and justifying decisions from the interpretation of data **(Ch. 23)**
*   solving problems that involve speed, distance and time **(Ch. 21)**
*   recording measurements using a scale on an instrument **(Ch. 22)**.

The Forth Road Bridge was opened in 1964.

It has a main span of 1006 m with two side spans measuring 408 m each.

The approach viaducts are 252 m long and 438 m long.

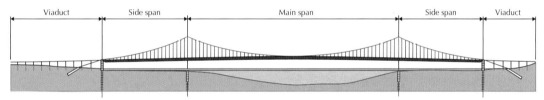

The graph shows the number of vehicles travelling on the Forth Road Bridge each year northbound (one way) from 1985 to 2007.

Number of crossings northbound on Forth Road Bridge

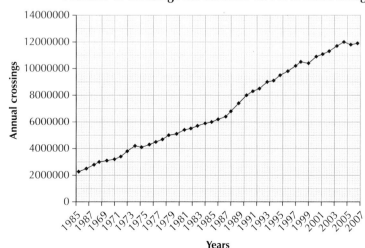

Source: transportscotland.gov.uk

325

The bridge was originally designed for no more than 11 million vehicles in a year.

In 1964, 2.5 million vehicles crossed the bridge.

In 2013, 22.9 million vehicles crossed the bridge.

The speed limit on the Forth Road Bridge is 50 miles per hour.

Sean is driving his car on the Forth Road Bridge. The dial shows his average speed.

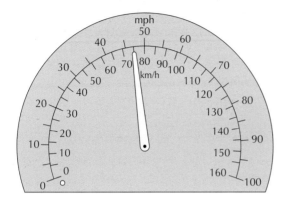

1 a Describe the trend in data.

[1C]

   b i   What percentage of the designed capacity was used in 1964?
      ii What was the percentage increase in vehicle crossings from 1964 to 2013?

[2P]

   c In 2005 the Scottish Government commissioned a new bridge to be opened in 2016.
      Give a reason for this decision.

[1C]

4 marks

2 a What is Sean's average speed in km/h?

[1 S]

   b How long does it take Sean to cross the Forth Road Bridge (including the approach
      viaducts)? Give your answer to the nearest second.

[1S, 2P]

   c Is Sean driving within the recommended speed limit? Give a reason for your answer.

[1C]

5 marks

Total 9 marks

Preparation for assessment: examination style questions (added value questions)

The final course assessment will be an examination (this is sometimes called the 'added value' unit). To prepare for this examination you should practise answering questions that require you to use added value skills and knowledge that go beyond the minimum competence that you met in the unit assessments. In this examination you will have to read and interpret real-life situations and then apply your skills and knowledge to solve problems from across the course. Practising on questions in past papers is a good way to prepare for your examination.

In this section there are examples to show how you should write your answers. There are also some added value questions for you to practise on.

When answering an examination question you should think about the following:

- *What maths can I use in this situation?* Identify the topic/concept, and then decide on a **strategy**.

- *What calculations do I need to do and how accurate do I need to be?* Show the **process** or working out that you are using and make sure your final answer make sense.

- *How should I* **communicate** *my answer?* Where possible relate your answer to the context in the question. Remember to use the correct units and rounding.

Example 1 (Paper 1, non-calculator)

Julie works in a coffee shop. The table below gives information about her monthly income and outgoings which can be used to work out her wage for the month.

Income		Outgoings	
Item	**Amount**	**Item**	**Amount**
Basic	£9.45 per hour	Cycle to work scheme	£20
Overtime	£15.80 per hour	Tax	£53.84
Tips	£58	NI Contribution	£11.20

If Julie works 40 hours basic and 5 hours overtime, calculate her net wage.

Process: Calculate basic and overtime

$£9.45 × 40 = £9.45 × 10 × 4$

$\qquad = £378$

$£15.80 × 5 = £15.80 × 10 ÷ 2$

$\qquad = £79$

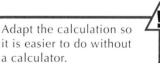

Adapt the calculation so it is easier to do without a calculator.

Strategy: Calculate net pay = total income − total outgoings

Process: Total income = £378 + £79 + £58

$\qquad\qquad = £515$

Total outgoings = £20 + £53.84 + £11.20

$\qquad\qquad = £85.04$

Net pay = £515 − £85.04

Communicate: Julie's net pay is £429.96.

Example 2 (Paper 2, calculator allowed)

Lonny bought a new car for £32 500 in August 2005. The value of his car falls by 20% in the first year and by 4% every year after that.

a Calculate the value of Lonny's car in August 2014. Give your answer to the nearest £100.

b The value of the average car falls by 5% a year. Has Lonny's car value fallen at the same rate as this trend? Give a reason for your answer.

a **Strategy:** Find value of car after 1 year (decrease of 20%)

Process: £32 500 × 0.80 = £26 000

Strategy: Find 4% for years 2–9

Process: £26 000 × 0.96^8 = £18 756.13 ≈ £18 800 (to nearest £100)

Communicate: Lonny's car is worth £18 800 in August 2014.

b **Strategy:** Find 5% fall over 9 years

Process: £32 500 × 0.95^9 = £20 483.11 ≈ £20 500 (to nearest £100)

Communicate: No, the value of Lonny's car (£18 800) is less than expected (£20 500).

Exercise

For each question, consider carefully your strategy, process and communication.

Paper 1 (Non-calculator)

1 Steven travels to the USA for a holiday. While there he buys a pair of jeans for $50.

    a If the exchange rate is $1 = £0.5970, what is the equivalent cost of the jeans in pounds?

    b On his return to Scotland Steven discovers that the exchange rate is now $1 = £0.6250. How many dollars would he have saved if he had bought the jeans at this new rate?

2 Mr Sexton has spent £835 this month using his credit card. Last month he had a balance of £238 owing. If he doesn't pay back anything on his card at the end of this month he will be charged the following:

    • £20 fee for non-payment

    • 2.5% monthly charge on the total outstanding.

Calculate the total he will have to repay if he doesn't pay anything this month.

3   Gillian is shopping for a new computer. She sees the same model in two different stores.

Which store should Gillian buy the computer from? Use your working to justify your answer.

4   John's grandfather uses a quick rule to roughly convert temperatures. To convert Fahrenheit to Celsius he subtracts 30 and halves the answer.

   a   Using this rule, what is the approximate temperature in Celsius if the thermometer reads 68°F?

   b   John tells his grandfather that last year the highest temperature in Scotland was 34°C. Describe how his grandfather will convert this temperature to Fahrenheit and calculate the temperature.

5   Matthew has 180 friends on a social media site. He notes down how often they post in a month and their age in a table. The results are shown below.

   If Matthew chooses a friend at random, what is the probability that:

   a   the friend was aged 16–19

   b   the friend was aged 20–29 and posted 21–40 times in that month?

Age (years)	Number of posts in a month		
	0–20	21–40	41–60
12–15	3	30	28
16–19	8	40	32
20–29	6	12	2
30–49	7	5	0
49+	4	3	0

6 Jack and Jill are taking part in an orienteering competition.

a Using the information in the table, draw an accurate scale drawing of the course using a scale of 1:20.

Point	Bearing from start	Distance from start (m)
A	122°	80
B	138°	140
C	171°	110
D	234°	35

b If a runner can maintain a speed of 4 km/h, how long will it take them to run from the start to A, B, C, D and return to the start?

7 Below is a diagram showing the design for an escalator to be built in a shopping centre. To meet European law, the following two points must be met:

1 the escalator must not move faster than 0.5 m/s

2 the escalator must not exceed a gradient of 1 in 1.428.

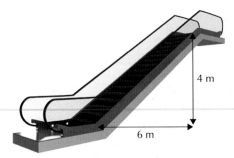

4 m

6 m

The escalator will travel approximately 6m in 14 seconds

a Does the escalator meet point 1? Give a reason.

b Does the escalator meet point 2? Give a reason.

c Does the design meet European law? Clearly state the reasons for your answer.

8 The table shows the distance travelled and the time taken by a delivery van for 10 deliveries in one day.

Distance (km)	9	25	42	26	45	32	38	30	13	19
Time (min)	25	71	130	75	142	90	115	93	40	58

a Draw a scatter diagram with time on the horizontal axis.

b Draw a best-fitting line on the diagram.

c Another delivery takes 1 hour.

   i How many kilometres would you expect the journey to have been?

   ii Calculate the average speed for this journey.

d How long would you expect a journey of 35 km to take?

9 A health centre has two doctors, Dr Collins and Dr Leckie. Data was collected to record the waiting time of patients and Dr Collins's results are shown below.

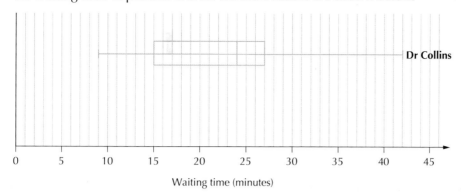

Waiting time (minutes)

a What is the median waiting time for Dr Collins's patients?

b Here is the data for Dr Leckie's patients' waiting times (in minutes). Construct a box plot to show this data. Use the same scale as shown in the box plot above.

10	6	15	30	21	18	21	20	32	10
13	9	22	21	15	29	16	19	12	8

c Make a valid comparison between the waiting times for Dr Collins's and Dr Leckie's patients.

10 Calculate the perimeter of a rectangular field $1\frac{1}{2}$ km long and $3\frac{2}{3}$ km wide.

11 Arriving at a T-junction on a cycle path Moira sees these road signs.

a At this point, what is the distance between Forres and Elgin?

b At the T-junction, how much further is it to Elgin than to Forres?

12 A plane leaves Inverness airport at 1125 and lands at Stornoway at 1205. Calculate the plane's average speed if the distance from Inverness to Stornoway is 120 miles.

13 A manufacturing company makes bespoke lamps for industrial units.

The table shows the list of tasks and the time required to complete each task.

Task	Detail	Preceding task	Time (hours)
A	Final design	None	5
B	Marketing input	A	4
C	Make base	B	2
D	Make upright section	B	3
E	Fit wiring	D	1
F	Fit on/off switch	D	0.5
G	Fit shade	C, E, F	0.5
H	Box up	G	0.25
I	Load onto vans for shipping	H	2

a Copy (enlarging your copy) and complete the diagram below by writing the tasks into the boxes to show the correct order.

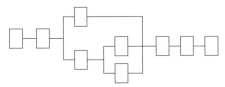

b The company claims the process of design to shipping is less than 16 hours. Is this true? Justify your answer.

c What is the critical path for this process?

14 Three friends jointly invest £18 000 in a new company. The ratio of the amount they invest is:

Aileen : David : Wendy = 4 : 7 : 9

a How much does each person invest?

b What percentage of the total amount has each person invested?

c The company makes a profit of £56 000. Profits are paid to each person in the same ratio as their investment. Wendy claims that she is due £30 800. Is she correct? Justify your answer.

15 The cross-section of a steel girder is shown in the diagram. Calculate the volume of the girder if its length is 2.5 m.

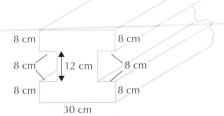

8 cm · 8 cm · 8 cm · 12 cm · 8 cm · 8 cm · 8 cm · 30 cm

Paper 2 (Calculator allowed)

1 A manufacturer of sweets called Happy Eats claims that the median contents of all its bags is 120 sweets.

A sample of six bags has the following number of sweets in each bag:

125, 127, 117, 113, 112, 130

a Does the data agree with the claims of the manufacturer?

A rival company produces a bag of sweets called Yummy Treats. The company states that every bag has 120 sweets ± 4.

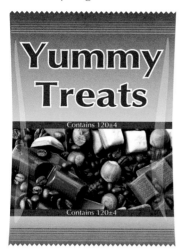

A sample of six bags has the following number of sweets in each bag:

120, 123, 123, 119, 116, 124

b Does the data agree with the claims of the manufacturer of Yummy Treats?

c Calculate the mean and standard deviation for each bag of sweets and make two valid comparisons about the number in each manufacturer's bag of sweets.

2 a Henry is a librarian who earns £20 140 per year. He has an annual personal allowance of £7970. He pays the basic rate of 20% tax.

He has to pay these deductions every month:

- £158 National Insurance
- £110 pension contributions
- £32 bike to work scheme.

Calculate Henry's monthly take-home pay.

b Kelley works for a gas installation firm. She is paid a basic wage of £1450 per month plus 3.5% commission on all sales. Her personal allowance is £9440.

i In January she sells £90 000 worth of equipment. Calculate her commission for January.

ii For the month of January she pays 40% tax and £252 NI. Calculate Kelley's take-home pay.

3 A boat is 3.5 km from Avoch on a bearing of 086° and a pod of dolphins is swimming 2 km from Fort George on a bearing of 328°.

a Make a scale drawing and plot the positions of the boat and the pod of dolphins.

b Write out a set of instructions for the boat's captain to sail to the dolphin pod. The boat must remain at least 100 metres from land at all times.

c The boat's captain wants to get to the dolphin pod as quickly as possible and sets off at 20 km/h. Can he reach the pod in less than 20 minutes? Justify your answer.

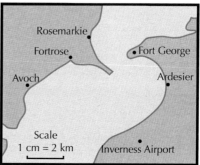

4 Sean, Hilda and Kevin are on holiday together in Ireland. They jointly buy some lottery tickets. They contribute the following amounts:

Sean €2, Hilda €12, Kevin €4

a They agree to share any winnings in the same ratio as their initial investment. State the ratio of investment in its simplest form.

b They win €23 463. How much does each person get?

c Kevin takes his winnings back to Scotland. Using the exchange rate of €1 = £0.7959, calculate how much money in pounds Kevin will receive.

d Sean decides to use his winnings to buy a new camera online. He finds a website selling a camera for JPY40 000 (Japanese yen). The exchange rate for converting pounds sterling to Japanese yen is £1 = JPY185.61. Does Sean have enough money to buy the camera? Justify your answer.

5 An ancient tunnel, cut from sandstone rock, has been found below the streets of Edinburgh. The cross-section of the tunnel is shown in the diagram. The top of the tunnel is a semicircle.

a If the tunnel is 50 m long, calculate the volume of rock that has been excavated to make the tunnel. Give your answer to 3 significant figures.

b The density of sandstone is 2323 kg/m³. Using the formula $D = \frac{m}{V}$, calculate the mass of the rock that was removed from the tunnel.

6 Two telegraph poles A and B are 4.5 m tall. They have wires connected to the roof of a building that is 23 m high. Pole A is 10 metres from the building.

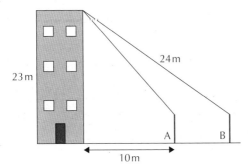

To the nearest centimetre, how far apart are the telegraph poles?

7 When constructing a thatch roof, best practice states that 'the pitch' or gradient of a thatched roof should be between 1 and $1\frac{3}{10}$. The weight of the thatch should not exceed 34 kg/m²'.

The diagrams below show the side profile of the roof and the front view of the roof. Does the rectangular part of the roof meet the requirements for best practice for thatched roofs? Justify your answer.

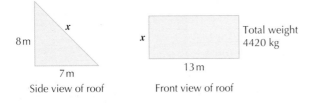

8 Mr and Mrs Cochrane would like to retile their roof. The table shows the three types they can choose from.

	Type of tile		
	Peg tiles	**Concrete tiles**	**Barrel tiles**
Life span	60 years	50 years	35 years
Number per square metre	70	60	40
Cost	£45 per 100	£0.55 each	£1.75 each
Labour costs	£50 per square metre	£0.30 per tile installed	£53 per square metre

A sketch of the roof of the Cochranes' house is shown below. The rectangular and triangular parts are to be tiled.

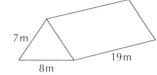

a Calculate the total cost of tiling the roof for each tile type.

b Which tile is the most cost-effective? Use your calculations to justify your answer.

9 Skye has been saving up to buy a new sofa. The same sofa is available in three different stores.

a Calculate the cost of each offer.

b Which is the best deal?

c Skye decides to borrow £480 from the bank to pay for the sofa. The loan will cost her 6.9% per annum until she starts to pay it back. If Skye pays nothing to the bank for 2 years, how much will she owe?

Extended case studies

Jo's case study

This case study focuses on the following skills:

- analysing a financial position using budget information **(Ch. 1)**
- analysing and interpreting factors affecting income **(Ch. 2)**
- determining the best deal, given three pieces of information **(Ch. 3)**
- selecting and using appropriate numerical notations **(Ch. 21)**.

Jo is a trainee electrician earning £16125 per annum.

Her personal tax allowance is £10000.

She pays tax at the basic rate of 20%.

In 2014–15 National Insurance is paid as follows:

Rate	Annual salary
0%	Less than £7956
12%	Between £7956 and £41860
2%	Above £41860

Jo is offered a room in a shared flat with three friends. To move into the flat her share of the monthly costs will be:

- rent £300
- energy costs £50
- council tax £60
- phone, broadband and television £15
- groceries £120.

Jo thinks that her friends are paying too much for their phone, internet and broadband service, so she researches different deals. The table shows the results of her research.

Company	Cost of TV	Number of channels	Cost of broadband	Cost of line rental	One-off set-up fee	Special offers
Optimum Online	£15.50	318	£8.25	£16.20	No upfront cost	First 6 months TV reduced to £8.20
RCT Communications	£14.75	82	£7.75	£16.25	£43.25	First 6 months broadband reduced to £5.50
Tower Technology	£11.50	80	£6.25	£16.45	£28	No offer

| Plug Telecoms | £10.95 | 125 | £7.60 | £16.50 | £44 | £30 supermarket gift card |
| Broadbandify Communications | £19.50 | 371 | £10.50 | £16.80 | No upfront cost | First 6 months broadband and TV reduced by 50% |

1 a Calculate Jo's monthly gross pay.   **[1P]**

   b How much income tax should she pay each month?   **[1P, 1S]**

   c How much National Insurance should she pay each month?   **[1P, 1S]**

   d Jo has joined her company's pension scheme. She pays 5.5% of her gross salary into the pension every month. Calculate how much she pays into her pension every month.
                                                                                                              [1P]

   e Using your answers to a–d, copy and complete the following payslip.

Martin's Electrix Ltd					
Name J McNab		**Employee number** 1284		**NI number** JH194727G	**Month** June
Payments	**Value**	**Deductions**	**Value**		
		Income tax NI Pension @ 5.5%	£ £ £		
Gross pay	£	**Total deductions**	£	**Net pay**	

                                                                                                              [1P]
                                                                                                        7 marks

2 Below is Jo's bank statement for June.

Date	Particulars	Paid out (£)	Paid in (£)	Balance (£)
05-Jun	Balance forward			900.25
07-Jun	BAC Car loan	200.00		700.25
08-Jun	DBT Petrol	55.23		645.02
08-Jun	DD Car service plan	30.00		615.02
13-Jun	DD Car insurance	65.00		550.02
15-Jun	DD Mobile phone	25.00		525.02
15-Jun	C/L Cash line	120.00		405.02
15-Jun	DBT Petrol	54.69		350.33
15-Jun	DBT Clothing store	40.00		310.33
29-Jun	BAC Wages			
30-Jun	Final balance			

a   What will Jo's final balance be?   [1P]

🦿 b   If her income and expenses remain the same every month, can Jo afford to
        move out of home and into the flat with her friends? Give a reason for
        your answer.   [1P, 1C]

                                                                                            3 marks

🦿 **3** a   By calculating the cost of the first year of each phone, internet and
              broadband service package, which company offers the best deal?   [1S, 1C]

🦿   b   If the special offers and set-up costs do not apply in the second year of
           each package, which company offers the best deal in this second year?

           Justify your answer.   [1C]

🦿   c   If they choose the company offering the best deal in the longer term
           (your answer to **b**), is this a better deal than their existing package?   [1C]

                                                                                            4 marks

                                                                                   Total 14 marks

The Zoo

This case study focuses on the following skills:

- planning a navigation course **(Ch. 12)**
- solving a problem using ratio **(Ch. 21)**
- interpreting scale diagrams **(Ch. 11)**
- solving problems using speed, distance and time **(Ch. 21)**.

Darius is a zoo keeper and has to feed the animals in part of a zoo.

The map shows the part of the zoo Darius is responsible for (including paths between animals).

Scale 1 : 4000

Darius can walk at a pace of 3 km/h.

He needs to spend 15 minutes at each type of animal to feed them.

He needs to have fed the penguins 15 minutes before their parade.

The penguin parade starts at 2.30 pm.

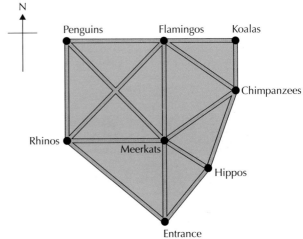

1   Darius has to visit all of the animals to feed them. He starts at the entrance and wants to finish at the penguins

    a   State the most efficient route for Darius to take.   **[2P, 1C]**

    b   Write instructions for Darius's journey by providing the distance and bearings for each part of the journey.   **[1S, 1P, 1C]**

                                       6 marks

2   **a**   What time does Darius need to start his journey in order to feed the penguins in time for the parade?   **[2S, 2P, 1C]**

    b   If it is raining Darius's average walking pace is reduced to 2.5 km/h and he now needs to allow 20 minutes at the hippos, penguins and flamingos. How much earlier will he have to start his journey on rainy days?   **[2P]**

                                       7 marks

                                  Total 13 marks

2011 Scottish Parliament Election

This case study focuses on the following skills:

- extracting and interpreting data from at least three different graphical forms **(Ch. 23)**

- solving problems by calculating percentages **(Ch. 21)**

- analysing and comparing two or more data sets using median, and semi-interquartile range **(Ch. 7)**

- explaining decisions based on the results of calculations **(Ch. 21)**.

Scottish Parliamentary Elections were held in Scotland in 2011.

1 991 051 people voted and there were 129 possible seats to be won by political parties.

Number of seats won

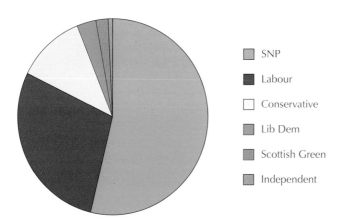

- SNP
- Labour
- Conservative
- Lib Dem
- Scottish Green
- Independent

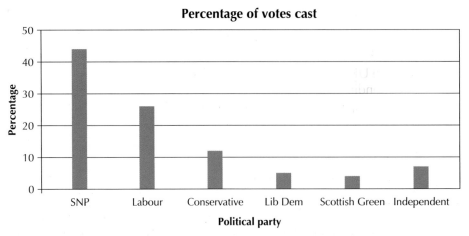

Percentage of votes cast

The back-to-back stem and leaf diagram shows the ages (in years) of SNP and Labour MSPs in 2011 at the time of their election.

Age of SNP MSPs		Age of Labour MSPs
9 6	2	6 7 8
9 8 4 2 1 1	3	0 4 5 7
9 9 9 9 8 8 8 7 7 7 5 4 3 3 3 3 2 2 1 1 1 1 0 0	4	0 2 4 5 7 8 8 8 8 9
9 9 8 8 8 8 7 7 6 5 5 5 5 4 4 4 3 3 3 3 2 2 1 1 0 0 0	5	0 1 3 4 4 5 5 7 7 7 8 9 9 9
9 4 2 1 1 1 0 0 0 0 0	6	0 1 1 2 2 9

$n = 69$ $6\,|\,2 = 26$ years $2\,|\,6 = 26$ years $n = 37$

The median of the Labour MSPs ages is 43 years and the semi-interquartile range is 7.75 years.

The Scottish Government spent a total of £53.1 billion in 2011.

Their spending can be split into different functions of the government: pensions, health care, education, welfare, with the remainder put into a category of 'other'.

This table shows the breakdown of expenditure in 2011.

Function	Spending (£ billion)
Pensions	12.7
Health care	10.8
Education	7.8
Welfare	
Other	13.8

1 a What percentage of votes did SNP win?   **[1S, 1P]**

  b What percentage of seats did Labour win?   **[1P]**

  c Comment on the percentage of seats and the percentage of votes won by the SNP and Labour Parties.   **[1C]**

                                                                     4 marks

2 a Calculate:

    i the median age of the SNP MSPs

    ii the semi-interquartile range of the age of the SNP MSPs.   **[1P, 1S, 1P]**

  b Compare the ages of SNP and Labour MSPs at the time of the election in 2011.   **[1C]**

                                                                      4 marks

3 **a** How much did the Scottish Government spend on welfare? **[1P]**

  b **i** What percentage of their budget did the Scottish Government spend on education? **[1P]**

    ii In the same year the UK Government spent £86.9 billion on education out of their total spending of £694.4 billion.

    Comment on the amount spent on education by the Scottish Government compared to the UK Government. **[1P, 1C]**

4 marks

Total 12 marks

Air travel

This case study focuses on the following skills:

- using a combination of statistical information presented in different diagrams **(Ch. 9)**
- using statistics to analyse and compare data sets **(Ch. 7)**
- solving problems involving speed, distance and time **(Ch. 21)**
- explaining decisions based on the results of calculations **(Ch. 21)**.

Henrik researches the cost of flights from Edinburgh to different cities in the UK (domestic flights) and the rest of the world (international flights).

The scatter graph shows the distance from Edinburgh and the costs of return flights for some domestic destinations.

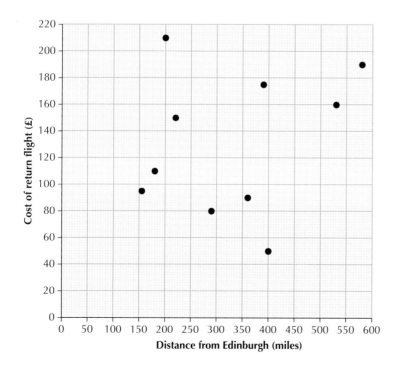

The table shows the distance from Edinburgh and the costs of return flights for some international destinations.

Destination	Cost of return flight (£)	Distance from Edinburgh (miles)
New York	600	3100
Amsterdam	120	740
Capetown	1100	6300
Athens	330	1750
Gran Canaria	350	2020
Dubai	640	3580
Reykjavik	200	819
Cairo	460	2450
St Petersburg	410	2020
Vancouver	760	4400
Mumbai	840	4600

Henrik researches the cost of a flight to New York using six different airlines.

His results are:

£600   £480   £390   £720   £520   £560.

Henrik chooses a flight to New York that leaves at 1415.

The aircraft average speed is 560 miles per hour.

The time difference is –5 hours.

1 a Using the data in the table above, create a scatter graph and draw a best-fitting line. **[1P, 1S]**

b The distance from Edinburgh to Bucharest is approximately 1500 miles.

Henrik finds a return flight from Edinburgh to Bucharest for £400. Is this a reasonable price? Give a reason for your answer. **[1P, 1C]**

c Compare the scattergraphs of domestic and international flights. What conclusion can you reach about the cost of flights and distance (the number of miles) to the destination? **[1C]**

5 marks

2 a Calculate the mean and standard deviation of the prices of flights to New York which Henrik has researched. **[1P, 1S, 1P]**

b The government raises the tax on the price of each flight to New York by £15. What is the new mean and standard deviation? **[1P, 1S]**

c Henrik also researches the cost of flights to Toronto. He finds the mean of six flights is £610 and the standard deviation is £12.50.

Make two valid comparisons between the prices of flights to Toronto and New York. **[1C]**

6 marks

3 a What time will Henrik arrive in New York (local time)? **[1S, 2P]**

b Henrik intends to go sight-seeing from 0900 to 1700 and expects to be in bed by 2300. If his mother wants to call him at his hotel, between what times should she call him? **[1C]**

4 marks

Total 15 marks

Removal day

This case study focuses on the following skills:

- solving a problem involving the volume of a composite solid **(Ch. 19)**
- using Pythagoras' theorem within a two-step calculation **(Ch. 20)**
- carrying out efficient container-packaging **(Ch. 13)**
- solving problems which involved speed, distance and time **(Ch. 21)**.

Jamie and Paula are moving house. They hire a removal van to make multiple journeys.

The storage space in the van is the shape of a cuboid and is 4 m long, 2 m wide and 2.3 m high. The weight of the load must not exceed 400 kg.

They use standard storage boxes. Each storage box has the dimensions shown.

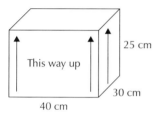

As part of the move, they need to take a pole that is 5.1 metres long.

They also have a concrete planter in the back garden which they want to take with them. It is a cylinder with a smaller cylinder removed from the centre as shown.

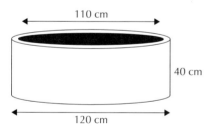

Concrete weighs 2.4 g per cm³.

They have eight loads to transport in the van from their old house to their new house. They can travel at an average speed of 40 km/h and the distance between the old house and the new house is 15 km. They estimate it will take them 30 minutes to load and unload the van each time.

The van hire company is 2 km from their old house and they can carry out this journey at an average speed of 25 km/h. The van hire company is 13 km from their new house and they can carry out this journey at an average speed of 42 km/h.

❀ **1** What is the maximum number of boxes that can be stored in the van?

Give a reason for your answer. **[1S, 1P, 1C]**
3 marks

❀ **2** Will the pole fit into the van? **[1S, 2P, 1C]**
4 marks

3   a   Calculate the volume of the concrete cylinder.   **[1S, 1P]**

     b   Can the concrete cylinder be transported in the back of the van? Give a reason for your answer.   **[1P, 1 C]**

         4 marks

4   They have hired the van for 14 hours. Is this enough time to move their belongings and return the van? Give a reason for your answer.   **[2P, 1C]**

         3 marks

         Total 14 marks

Fundraising day

> **This case study focuses on the following skills:**
>
> - solving problems that involve perimeter and volume **(Ch. 18, 19)**
>
> - solving problems that involve ratio and proportion **(Ch. 21)**
>
> - explaining decisions based on the result of calculations **(Ch. 21)**
>
> - making use of probability and expected frequency to investigate risk **(Ch. 24)**.

Neeta bakes cakes to sell at a local fundraising day.

The list shows the ingredients for a vanilla sponge cake for 6 people.

The cake tin she uses is in the shape of a cylinder and has a diameter of 24 cm.

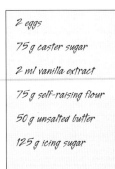

2 eggs

75 g caster sugar

2 ml vanilla extract

75 g self-raising flour

50 g unsalted butter

125 g icing sugar

Joanne is running a game at the fundraising day. In the game the player rolls two balls down the sloping board. The player wins a prize of £2 if the balls land in slots that total more than 7.

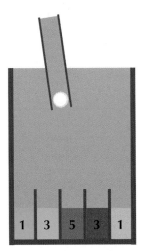

1   a   How much of each ingredient is needed to make a cake for 10 people?   **[1S, 1P]**

     b   If the cake reaches a height of 8 cm, what is the volume of the cake? Give your answer correct to 3 significant figures.   **[1S, 1P]**

c What length of ribbon is required to go round 10 cakes? Allow 1 cm for overlap. **[1P]**

d If the same amount of cake mixture is put into a square tin with side length 20 cm, what height do you expect it to rise to? Give your answer correct to 3 significant figures. **[1P, 1C]**

e i How much ribbon is required to go round 10 cakes with a square base? Allow 1 cm for overlap.

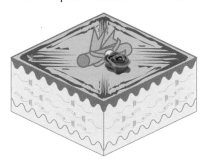

[1P]

ii Based on the amount of ribbon required, which cake is better value for money, the cylindrical cake or the cake with the square base? Justify your answer

[1C]

9 marks

2 In the game Joanne is running calculate the probability of rolling more than a seven. Use the probability tree diagram below to help you.

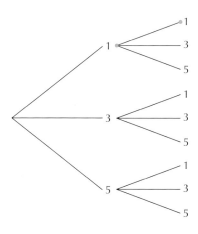

[1S, 2P]

b i If 200 people play the game, how many would you expect to win?

ii If Joanne charges 50p per go, how much money does she expect to make if 200 people play?

[1S, 1P]

5 marks

Total 14 marks